SpringerBriefs in Earth Sciences

For further volumes:
http://www.springer.com/series/8897

D. Fraser Keppie

The Analysis of Diffuse Triple Junction Zones in Plate Tectonics and the Pirate Model of Western Caribbean Tectonics

 Springer

D. Fraser Keppie
Petroleum Resources
Nova Scotia Department of Energy
Halifax
Canada

ISSN 2191-5369   ISSN 2191-5377 (electronic)
ISBN 978-1-4614-9615-1   ISBN 978-1-4614-9616-8 (eBook)
DOI 10.1007/978-1-4614-9616-8
Springer New York Heidelberg Dordrecht London

Library of Congress Control Number: 2013955879

Printed on acid-free paper

Springer is part of Springer Science+Business Media (www.springer.com)

Acknowledgments

I would like to acknowledge the generosity, guidance, and support of Dr. Michael Gurnis and the California Institute of Technology for supporting this work during a postdoctoral appointment. I thank Dr. Rob Clayton, Dr. Joann Stock, Dr. Andrew Hynes, Dr. Steven Skinner, Dr. Brendan Murphy, Dr. J Duncan Keppie, and Dr. Keith James for their reviews of the manuscript. I thank Dr. Lisa Gahagan, Dr. Rob Rogers, and Dr. Ian Norton for providing the rotation table from the UTIG Plates project for relative North American-Caribbean plate motion based on Leroy et al. (2000). I also thank developers responsible for the Generic Mapping Tools (GMT), GPlates, and Quantum GIS (QGIS) software projects used in the preparation of this manuscript.

Contents

Abbreviations

B	Extension
C	Shortening
Ca	Caribbean
CAMA	Central American magmatic arc
Co	Cocos
D	Displacement
E	East
F	Fault
FFT	Fault-Fault-Trench
h	Distance
MAT	Middle America Trench
N	North
Na	Nazca
NA	North America
NCB	North Caribbean Boundary
R	Ridge
S	South
SA	South America
SMT	South Mexico Trench
T	Trench
TTT	Trench-Trench-Trench
UTIG	University of Texas, Institute for Geophysics
W	West

Chapter 1
Introduction

Abstract The tectonic evolution of the Caribbean region since the late Cretaceous is of fundamental significance to plate tectonics. Heretofore, widely-accepted geological interpretations have held that the east-dipping Middle America subduction zone at the west margin of the Caribbean region and the west-dipping Lesser Antilles subduction at the east margin of the Caribbean region have remained stationary relative to one another since ca. 70 Ma or earlier. In this way, a single rigid Caribbean Plate is thought to have occupied the whole Caribbean region following the initiation of subduction at both the Middle American and Lesser Antillean trenches. Modern geographic positioning system (GPS) data and geomorphological displacements indicate that non-rigid processes have been important for the tectonic evolution of the Caribbean region. However, these neo-tectonic results have not modified the common view that the width of the Caribbean Plate has remained essentially constant throughout the Cenozoic.

The consequence of a rigid Caribbean Plate of constant width is that the rates of slab-hinge retreat or trench rollback of the two bounding subduction zones cannot have been controlled by lower plate forces such as slab pull or slab suction. This is because the subducting slabs belong to completely different lower plates at the western and eastern subduction zones. Pacific lithosphere from the ancient Farallon Plate has subducted under the western Middle America Trench, whereas Atlantic lithosphere from the modern North and South American Plates has subducted under the eastern Lesser Antilles Trench. If lower plate mechanisms were important, different proportions of slab pull or slab suction due to the different proportions and character of subducted lithosphere at the respective trenches would be expected to involve different rates of slab-hinge retreat at the bounding subduction zones and thus a change in the width of the Caribbean Plate through time. Therefore, short of an extraordinary coincidence, lower plate mechanisms do not appear to be important in determining either slab-hinge or trench retreat rates of the subduction zones bounding the Caribbean Plate.

This result is surprising because it contradicts the emerging view that global plate tectonics on Earth may be driven by a subduction-derived or top-down driving mechanism, in general. For example, the Pacific Ocean region has been shrinking in conjunction with the progressive widening of the Atlantic Ocean region since the breakup of Pangea—and this has been occurring in spite of the fact that the Pacific Ocean is rifting at a faster rate than the Atlantic Ocean. The new realization is that the partitioning of negative buoyancy forces produced by sinking slabs may explain

D. Fraser Keppie, *The Analysis of Diffuse Triple Junction Zones in Plate Tectonics and the Pirate Model of Western Caribbean Tectonics,* SpringerBriefs in Earth Sciences, DOI 10.1007/978-1-4614-9616-8_1, © Springer Science+Business Media New York 2014

this global tectonics most simply. Partitioning of downward slab forces would lead to both lower plate advance relative to a stationary mantle and slab foundering with associated slab-hinge retreat relative to a stationary upper plate. These dual effects of slab sinking provide simultaneous and complementary explanations for rifting within the fast-spreading Pacific and slow-spreading Atlantic realm. Such interpretations support the idea that plate tectonic systems evolve due to subduction-derived or top-down forcing, but, as just noted, the inferred existence of a long-lived Caribbean Plate of constant width is contradictory to this.

The tectonic evolution of the Caribbean region since the late Cretaceous is of fundamental importance at least in part because it may preserve a key counter-example to the subduction-driven or top-down plate tectonics inferred elsewhere. This important prospect has questionable merit at the present time, however, because the standard reconstructions of Caribbean tectonics while widely-accepted are poorly-tested. The present work attempts to address this shortcoming by identifying a common theoretical framework with which to compare the predictions of the competing models of Caribbean tectonics. Only through a consideration of the alternatives can confidence be gained for a preferred solution. In the present context, the consideration of alternative interpretations of the Caribbean rock record is prerequisite to attaining confidence that the Caribbean region is indeed globally anomalous with respect to the driving mechanisms of plate tectonics.

Although the evolution of the Caribbean Plate (Fig. 1.1) has been studied for over half a century, and the general consensus is that it originated in the Pacific Ocean (Pindell and Dewey 1982; Burke 1988; Ross and Scotese 1988; Mann 2007; Pindell and Kennan 2009; Torres de Leon et al. 2012; Talavera-Mendoza et al. 2013), a few papers have raised serious questions (Molnar and Sykes 1969; Burkart and Self 1985; Guzmán-Speziale et al. 1989; Ego et al. 1995; Ego and Ansan 2002; Audemard and Audemard 2002; Keppie and Morán-Zenteno 2005; James 2006; Morgan et al. 2008; James 2009; Guzmán-Speziale 2010; Keppie 2012; James 2013). The most serious question is the absence of fault connections between the north and south faults bounding the Caribbean Plate, and the Middle America Trench (Molnar and Sykes 1969; Meschede and Frisch 1998; Guzmán-Speziale et al. 1989; Audemard et al. 2005; Authemayou et al. 2011; Fig. 1.2). Such fault connections are an essential element for a Pacific origin (Fig. 1.3a; Keppie and Morán-Zenteno 2005; James 2006; Morgan et al. 2008). This has led (James 2006, 2009, 2013) to resurrect the possibility of an in situ origin (Fig. 1.3b; Ball et al. 1969). However, an in situ origin requires little or no relative motion across the north and south bounding faults (Pindell et al. 2006), which is at odds with the ca. 1,100 km or greater strike-slip displacements reported by various authors (Rosencrantz and Sclater 1986; Leroy et al. 2000; Mann 2007; Escalona and Mann 2011).

In this volume, I apply a new approach to the question of western Caribbean tectonics (Chap. 2). I first consider western Caribbean tectonics from a theoretical perspective and then I compare the competing theoretical predictions of different end-member model possibilities with the geological record. This approach

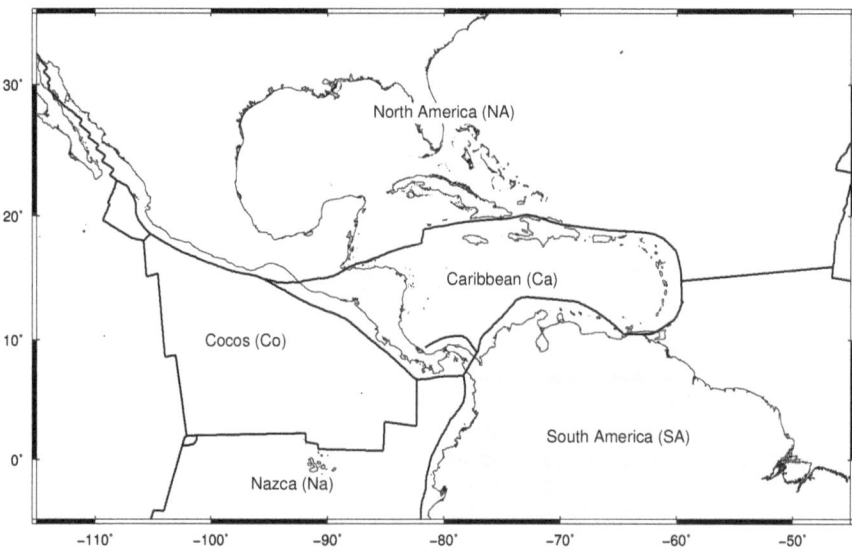

Fig. 1.1 Major plates with simplified boundaries for the western Caribbean region (after ESRI 2011) showing North America (*NA*), South America (*SA*), Caribbean (*Ca*), Cocos (*Co*), and Nazca (*Na*) plates

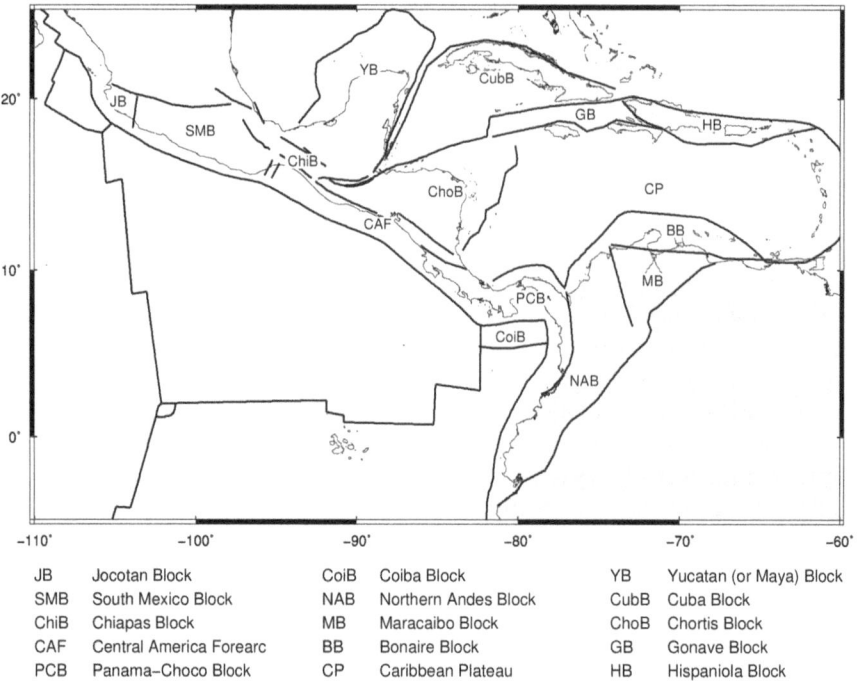

JB	Jocotan Block	CoiB	Coiba Block	YB	Yucatan (or Maya) Block
SMB	South Mexico Block	NAB	Northern Andes Block	CubB	Cuba Block
ChiB	Chiapas Block	MB	Maracaibo Block	ChoB	Chortis Block
CAF	Central America Forearc	BB	Bonaire Block	GB	Gonave Block
PCB	Panama–Choco Block	CP	Caribbean Plateau	HB	Hispaniola Block

Fig. 1.2 Fifteen microplates with simplified boundaries for the western Caribbean region (based on various publications cited in the text)

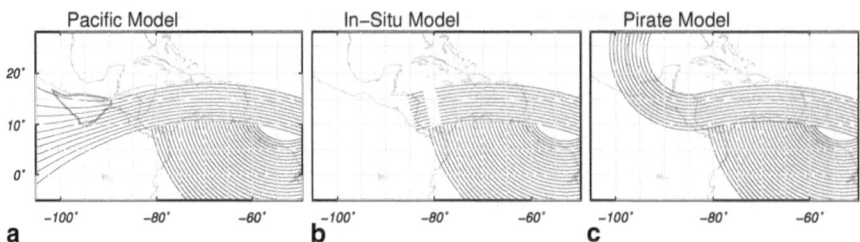

Fig. 1.3 End-member model types for the tectonic evolution of the western Caribbean region. Representative material flow lines are plotted to show the general inferred motion(s) for western Caribbean lithosphere through time: **a** Pacific model, in which *flow lines* show the inferred motion of western Caribbean lithosphere from west to east and the inferred narrowing of Caribbean lithosphere in a north–south direction as the Chortis Block overrides the Caribbean Plateau, **b** In-Situ model, in which *flow lines* terminate in zones of crustal thinning and extension and (possibly) magmatic intrusion in the west and/or central Caribbean regions, and **c** Pirate model, in which *flow lines* show the inferred motion of western Caribbean lithosphere captured out of the eastern Mexico/western Gulf of Mexico region. All models incorporate the documented clockwise flow of South American lithosphere into the southwest Caribbean region as shown. However, clockwise flow is only a predicted feature of Pirate models of Caribbean tectonics wherein it is viewed as lithospheric capture at the scale of a major plate

allows evaluation of the Pacific and in situ models, but more importantly permits a new, Pirate model for the plate tectonic evolution of the western Caribbean region (Keppie 2012, 2013). The Pirate model occurs when plate material from North and South America rotates into the trailing edge of the Caribbean Plate about vertical axes proximal to the Caribbean Plate (Fig. 1.3c; Keppie 2012; Keppie and Keppie 2012). Counter-clockwise rotation of northwest Caribbean lithosphere (i.e., the Chortis Block, Fig. 1.2) out of eastern Mexico/western Gulf of Mexico and clockwise rotation of southwest Caribbean lithosphere out of northern South America are the primary predictions of the Pirate model (Fig. 1.3c).

Comparison with the geological record suggests that all of the different end-member models have been important for different periods of geological history. In this volume, I conclude that the operation of the Pacific and in situ models may have been most important in the early evolution of the western Caribbean region (from ca. 120 to 80–55 Ma), but that the Pirate model may have dominated its later evolution (from ca. 80–55 to 10–0 Ma) once unstable, underlying plate boundary triple junction configurations formed at western Caribbean Plate corner zones.

The new, Pirate model has additional regional implications for the timing and cause of a number of western Caribbean tectonic phenomena previously poorly explained and generally considered in isolation from one another (Keppie 2012). Examples of these regional implications include: (1) the extension, subsidence, and counter-clockwise opening in the Gulf of Mexico (e.g., Dickinson 2009) which may have been related in part to the capture of the Chortis Block from the gulf's western part (Keppie 2012), (2) the paleogeography of the Chortis block (e.g., Keppie and Morán-Zenteno 2005) which may be allocthonous but derived from eastern, rather than southern Mexico (Keppie 2012), (3) the neotectonic (and possibly older) dextral transtension observed in the Central American arc/backarc region (e.g., La Femina et al. 2002) which may be related to the counter-clockwise capture of microplates from southern North America (Authemayou et al. 2011; Keppie 2012), (4) sinistral shear inferred along the eastern margin of the Choco Block (e.g., Ego et al. 1995) which may be related to the clockwise capture of the South American Plate and smaller microplates from the northern margin of South America, and (5) the southward subduction of the Caribbean Plate under northwest South America (e.g., James 2013) which may reflect the convergence of western Caribbean lithosphere entering the Central American arc/backarc region from the north and south at the same time (Keppie 2012, 2013).

The theoretical analysis used here is explained in greater detail in Chap. 3; the initial result of this analysis is depicted for both western Caribbean corner zones in Fig. 1.4, which I now explain. Basically, the major plate motions and the major plate boundaries (e.g., Fig. 1.1) provide the context for the simple or complex evolution of the western Caribbean region. All of the plate boundaries between these major plates are known to be complicated during the Cenozoic by diffuse deformation patterns that appear to be both heterogeneous and time-varying (e.g., Mann 2007; Audemard 2009; Demets et al. 2010; Authemayou et al. 2011); even the microplate boundaries depicted in Fig. 1.2—which attempt to express some of the complexity—oversimplify the distributed deformation reported across these boundaries for different times. However, when studies document and explain these complexities directly and locally (e.g., Pindell and Dewey 1982; Burke 1988; La Femina et al. 2002; Andreani et al. 2008; Audemard 2009; Pindell and Kennan 2009; Authemayou et al. 2011; James 2013), with the full complexity intact, it can be difficult to consider whether properties of the major plate system may in fact explain some or all of the local deformation patterns (e.g., Keppie 2012, 2013). The method I explain in Chap. 3 and apply here attempts to overcome this difficulty.

Chapter 3 explains a normalization analysis in which four steps are performed: (1) the selection of a moving tangent tectonic reference frame in which Euclidean analysis of vector displacements and velocity changes can be undertaken both locally and along material flow lines, (2) the approximation of complex major plate boundary zones as infinitesimally broad classic plate boundary types, (3) the analysis of hypothetical triple junction configurations formed by the normalized plate boundary zones according to a published geometric method (McKenzie and Morgan 1969), and (4) if the hypothetical triple junctions are unstable, construction of the end-member models of further system evolution which could resolve the underlying instabilities.

Figure 1.4 depicts the result of the first two steps of this procedure applied to the western Caribbean region. The major plate instabilities underlying the evolution of both western Caribbean corner regions (calculated explicitly in step 3 of the normalization analysis) is transparent because the northern and southern Caribbean plate boundary zones are seen to strike toward the continuous western American subduction system at a high angle (Fig. 1.4; Morgan et al. 2008; Authemayou et al. 2011; Keppie 2013). Further evolution of such normalized triple junctions require some sort of nonrigid deformation to take place (Chap. 3). In general, the nonrigid

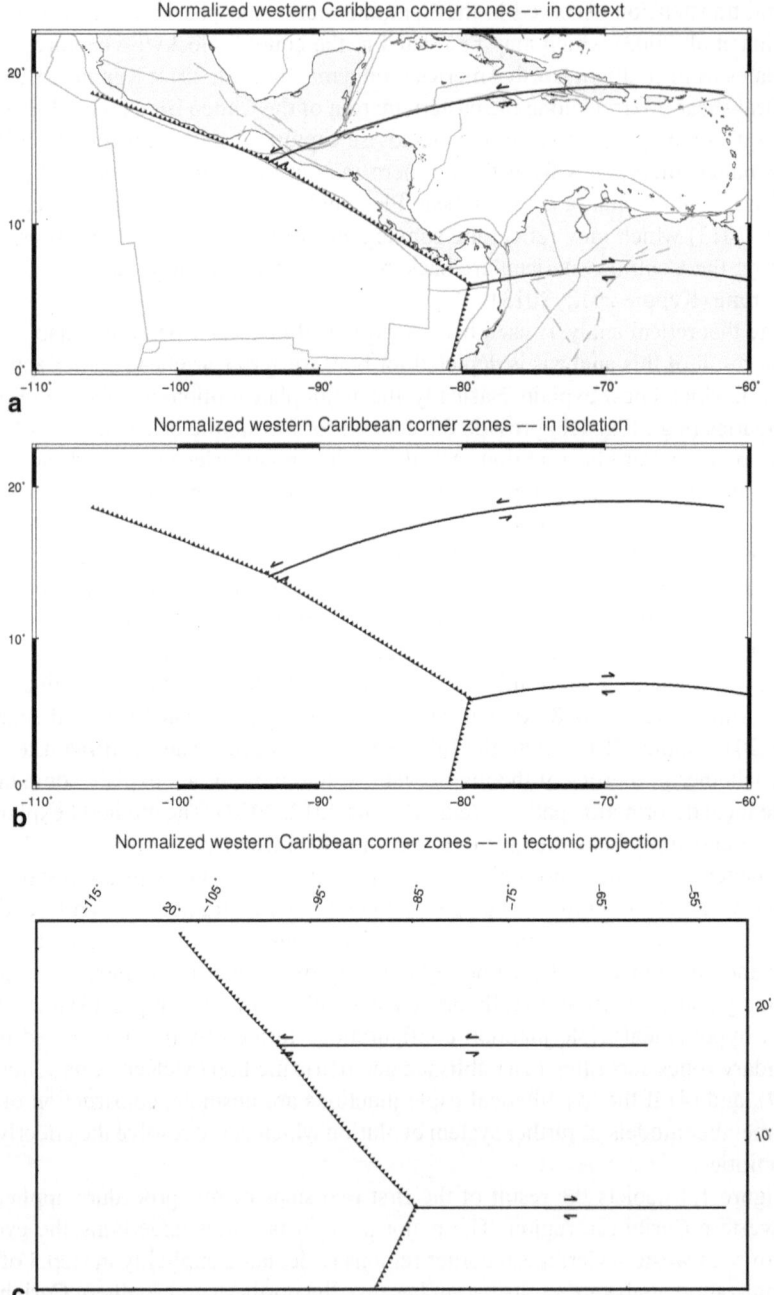

Fig. 1.4 Simplified view of hypothetical western Caribbean corner triple junctions following the normalization analysis outlined in Chap. 3. Normalized triple junctions at the western Caribbean Plate corners are unstable because the northern and southern Caribbean plate boundaries strike at an angle to the western American subduction zone segments. Normalization of the northern

deformation possibilities include: plate amalgamation events, plate breakup events, plate velocity change events, or the emergence of steady-state lateral tectonics (Chap. 3; Keppie 2012, 2013). In this volume, however, I consider various possibilities for the past (and present) evolution of the apparently unstable (hypothetical) western Caribbean corner zones (Fig. 1.4) in terms of the three end-member models proposed for the evolution of the western Caribbean region (Fig. 1.3). In Fig. 1.5, I show some of these possibilities and my association with the different models of western Caribbean tectonics. I discuss Fig. 1.5 in detail further in Chap. 2.

One insight that can be brought to bear on western Caribbean tectonic analysis is that all of the proposed end-member models are theoretically valid and none are mutually exclusive (Keppie 2013). This insight can be made transparent by comparing the end-member western Caribbean models with the end-member ways mass can be conserved (incompressibly) in the western Caribbean region (Fig. 1.6). If the main part of the Caribbean Plate is exiting the western Caribbean region (relative to North and South America), new lithosphere must enter the western Caribbean region from somewhere lest a big hole open up in the surface of the Earth. Basically, there are only three directions from which material can enter the western Caribbean region: from the west, from depth, or from the north and south which correspond directly to the x, z, and y directions of the moving tangent Euclidean reference frame chosen here in which to analyze western Caribbean tectonics (Chap. 3). If one finds moving tangent Euclidean reference frames an unfamiliar concept, it is possible to simply view the x, z, and y directions discussed herein to be nominally equivalent to projected latitude, depth, and projected longitude in a tectonic coordinate system constructed relative to the net rotation between North America and the Caribbean since 47.9 Ma. If the concept of tectonic coordinate systems is also unfamiliar, it is possible to simply view the x, z, and y directions discussed herein to correspond approximately to conventional geographic latitude, depth, and longitude (at least local to the western Caribbean region).

Caribbean Plate boundary zone and northwest Caribbean corner zone follows the method given in Chap. 3 exactly. Normalization of the southern Caribbean Plate boundary zone and southwest Caribbean corner zone follows a slightly modified implementation of this method. The infinitesimally wide flow line chosen to approximate the southern Caribbean Plate boundary zone is taken as an NA-Ca net tectonic flow line corresponding roughly to the paleo-position at 55.9 Ma of the dextral fault system running along the northern margin of the Maracaibo block (refer to *dashed grey lines* in *a* which reflect these and a few other northern South American faults back-rotated to 55.9 Ma after the model of Seton et al. 2012). Modelling the South American boundary in this way allows us to identify the implied southwest Caribbean corner triple junction configuration prior to clockwise rotation of South America relative to North America and to use the same moving, tangent reference across the whole Caribbean region for triple junction analyses and comparison of velocities and displacements calculated at different points along the various material flow lines hypothesized in the different western Caribbean tectonic models. **a** The normalized western Caribbean corner triple junctions with continental and plate boundaries included to provide context. **b** The same configurations without the context which sets up the qualitative, abstract discussion undertaken in the first part of Chapter 2. **c** The same configurations reprojected in terms of the NA-Ca tectonic coordinate system introduced in Chap. 3 to emphasize the Euclidean perspective with which one can view this reference frame

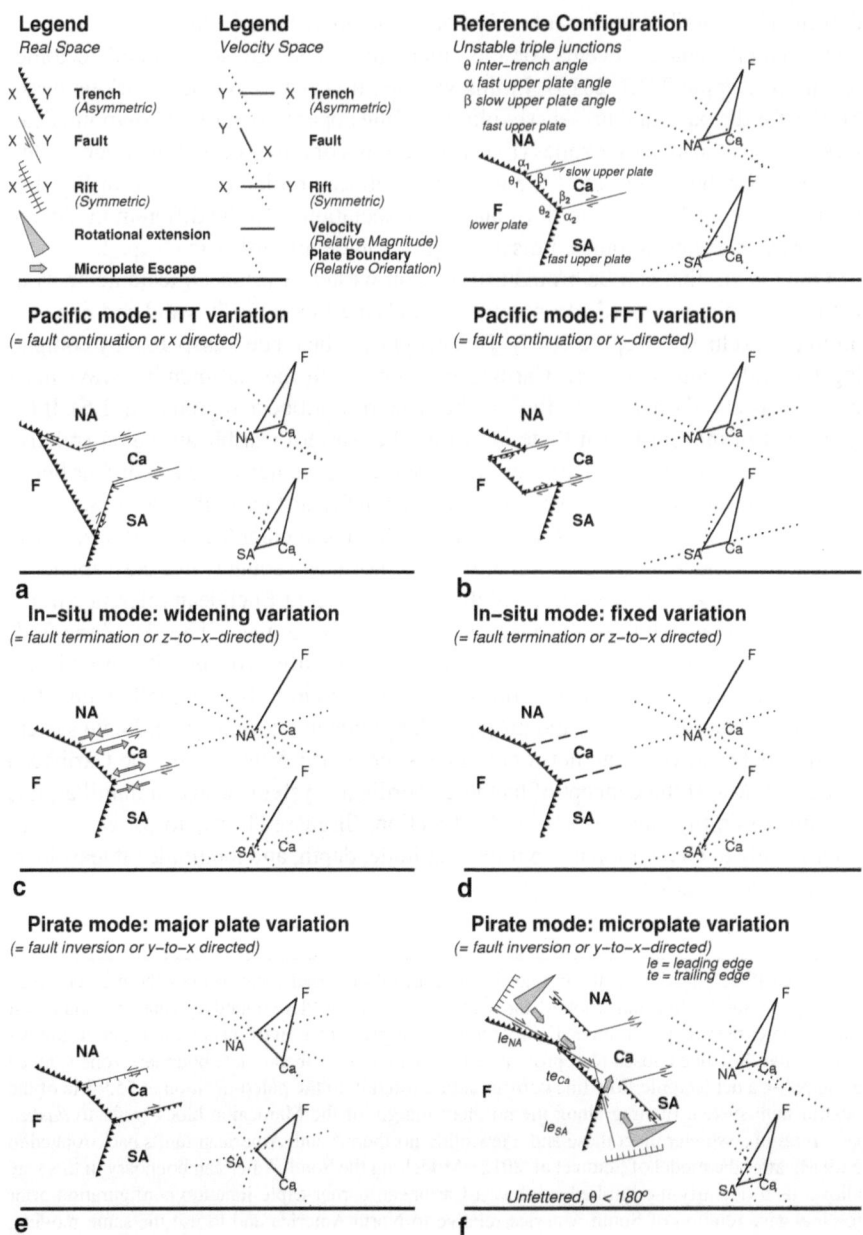

Fig. 1.5 Various simplified, theoretical possibilities for the tectonic evolution of the western Caribbean region. **a** Pacific model, TTT-variation. **b** Pacific model, FFT-variation. **c** In-situ model, widening-variation. **d** In-situ model, fixed-variation. **e** Pirate model, major plate-variation. **f** Pirate model, microplate variation

Relationship between western Caribbean tectonic model types and mass conservation modes

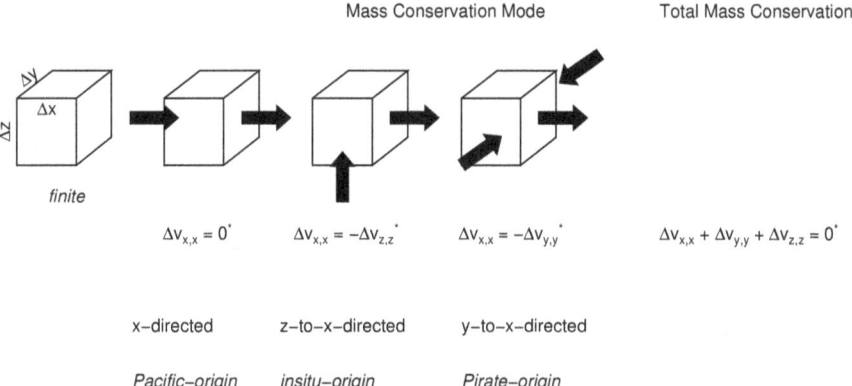

Mass Conservation Mode   Total Mass Conservation

$\Delta v_{x,x} = 0^*$ $\Delta v_{x,x} = -\Delta v_{z,z}^*$ $\Delta v_{x,x} = -\Delta v_{y,y}^*$ $\Delta v_{x,x} + \Delta v_{y,y} + \Delta v_{z,z} = 0^*$

x–directed   z–to–x–directed   y–to–x–directed

Pacific–origin   *insitu–origin*   *Pirate–origin*

Fig. 1.6 The conceptual relationship between the competing models of western Caribbean tectonics (Pacific, In-situ, and Pirate) and the theoretical kinematic constraint supplied by the assumption of incompressible mass conservation across finite regions of the Earth's surface. * The notation $v_{x,x}$ refers to the x-directed spatial derivative of the x-directed velocity component; similar definitions apply likewise for $v_{y,y}$ and $v_{z,z}$

One significance of noting that each end-member model proposed for western Caribbean tectonics (Fig. 1.3) satisfies end-member models of incompressible mass conservation (Fig. 1.6) is that fundamental geodynamic modeling may struggle to discriminate between which model is more likely. Rheologies can be specified for the main plates and presumed faults to facilitate the theoretical evolution of the western Caribbean region in terms of any of the kinematic proposals. Further, the evolution of the western Caribbean region may incorporate components of all three end-member possibilities at the same time or in sequence over its full history. Detailed reading of the different proposals for modern Pacific models (e.g., Pindell and Kennan 2009), modern in situ models (e.g., James 2009), and the new Pirate model (Keppie 2012 and this volume) will show that each account incorporates aspects of the others in modest amounts. However, substantial differences remain with regard to the interpretation of timing and dominant evolutionary model of the western Caribbean region, which motivates the present study.

In the present study, I first review the theoretical evolutionary possibilities for the western Caribbean region given that its western corner zones appear unstable at present when normalized in terms of classic plate tectonic boundary types. As noted, three theoretical categories of model are forthcoming when such an approach is taken (Figs. 1.3 and 1.6). Each of the theoretical models carries implications for the interpretation of the western Caribbean tectonic record, but not all are similar, not all are unique, and, as noted, not all are mutually exclusive. Thus, later, in order to evaluate the possible relevance a documented phenomena may have, I plot the reviewed phenomena in up to 3-member Venn diagrams (Fig. 1.7). This allows a quick determination of the phenomena which is diagnostic of a specific model

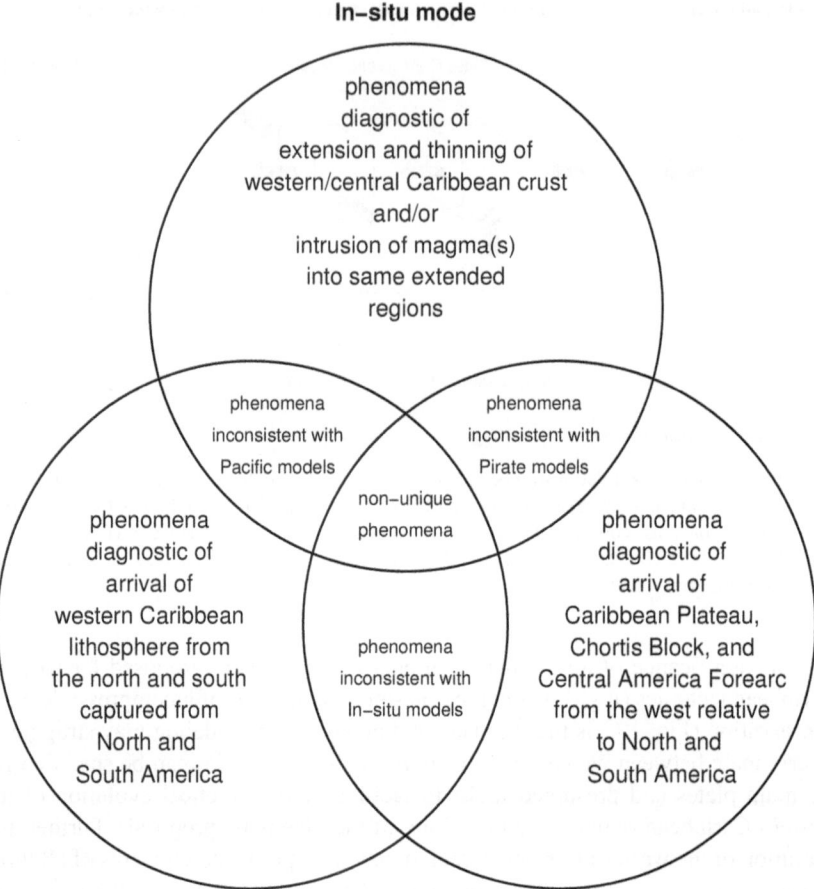

Fig. 1.7 A three-member Venn diagram appropriate for the general comparison of western Caribbean tectonic phenomena in terms of the three competing models of western Caribbean tectonics. In this framework, documented phenomena can be inferred to be diagnostic of a given end-member, a diagnostic against a given end-member, or common to all models and not useful for model discrimination purposes

versus the others, diagnostic in the sense that it excludes one of the three end-member models versus the others, or is nonunique in that it is consistent with all three of the end-member models. Using this approach, it is possible to compare not only the predictive power of each end-member model, but also where phenomena exists in the absence of a prediction from a given model.

Chapter 2
Western Caribbean Tectonics

Abstract Despite 50 years of study, the evolution of the western Caribbean Plate region is still debated and there are three possible end-members: (1) Pacific model where the western Caribbean lithosphere is derived from the eastern Pacific and the northern and southern Caribbean Plate boundaries connect directly west to the Middle America Trench at the western Caribbean Plate boundary, (2) In-situ model where the western Caribbean lithosphere is derived from depth and the northern and southern Caribbean Plate boundaries terminate in a broad zone of extension in the western Caribbean Plate, and (3) Pirate model where the western Caribbean lithosphere is derived from the southern and northern margins of North and South America, and the northern and southern Caribbean Plate boundaries have either accommodated convergence themselves, or have curved to the north and south prior to reaching the Middle America Trench. Analysis indicates all models have been important for the evolution of the western Caribbean at different times but the Pirate model may have been dominant during the Cenozoic. The Pirate model resolves the absence of fault connections between the northern and southern boundaries of the Caribbean Plate with the Middle America Trench that are essential for the Pacific model and the > 1,100 km of net strike-slip displacements inferred across the northern and southern Caribbean margins that are unexplained by the in-situ model. In the Pirate model, North and South American material is inferred to have rotated into the trailing edge of the Caribbean Plate across the western Caribbean Plate corners.

2.1 Tectonic Constraints

Western Caribbean tectonics is an ideal setting to consider the theoretical implications of mass conservation and triple junction stability in because of its unique plate boundary conditions (Chapter 1). The key factors relate to opposed subduction zones, which are east-dipping at the western boundary and west-dipping at the eastern boundary of the Caribbean Plate, and to the longer western than eastern boundary subduction zone. This creates a situation where the northern and southern Caribbean Plate boundaries strike at an angle to the western trench and imply unstable triple junctions at the western Caribbean plate corners (Fig. 1.4, Chap. 3). The purpose of this chapter is to evaluate how the western Caribbean system may be

D. Fraser Keppie, *The Analysis of Diffuse Triple Junction Zones in Plate Tectonics and the*   11
Pirate Model of Western Caribbean Tectonics, SpringerBriefs in Earth Sciences,
DOI 10.1007/978-1-4614-9616-8_2, © Springer Science+Business Media New York 2014

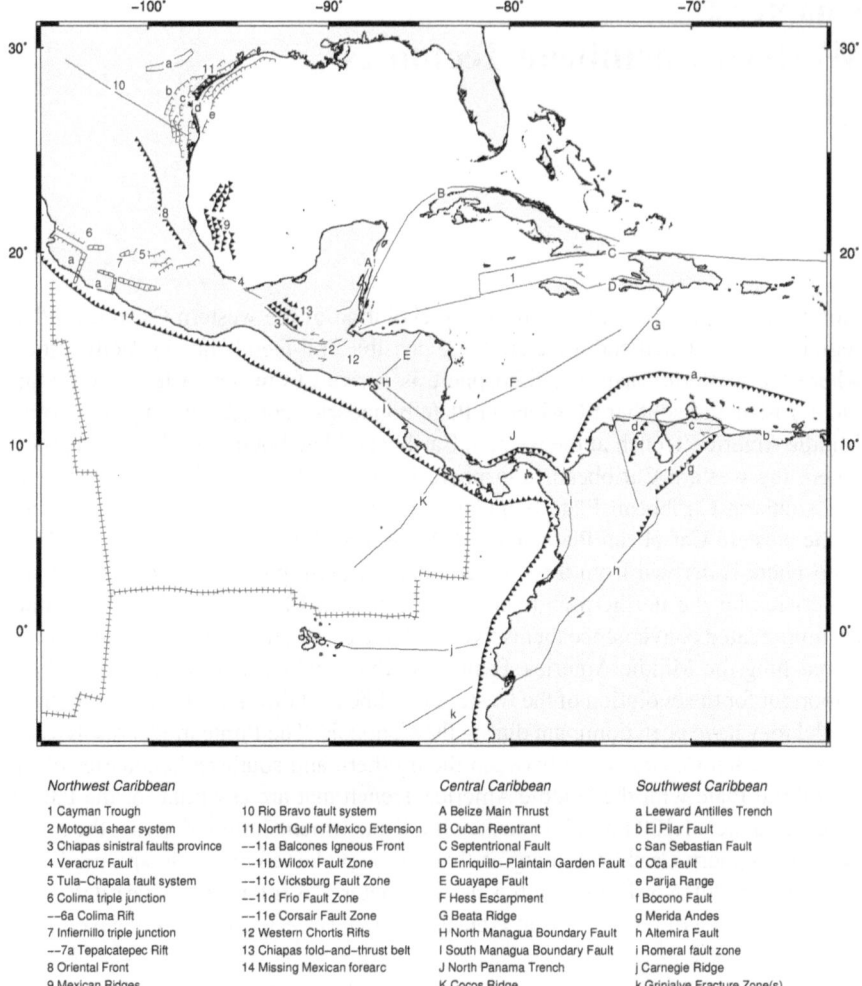

Fig. 2.1 Tectonic features in the greater western Caribbean region (see text for references)

Northwest Caribbean		Central Caribbean	Southwest Caribbean
1 Cayman Trough	10 Rio Bravo fault system	A Belize Main Thrust	a Leeward Antilles Trench
2 Motogua shear system	11 North Gulf of Mexico Extension	B Cuban Reentrant	b El Pilar Fault
3 Chiapas sinistral faults province	--11a Balcones Igneous Front	C Septentrional Fault	c San Sebastian Fault
4 Veracruz Fault	--11b Wilcox Fault Zone	D Enriquillo–Plaintain Garden Fault	d Oca Fault
5 Tula–Chapala fault system	--11c Vicksburg Fault Zone	E Guayape Fault	e Parija Range
6 Colima triple junction	--11d Frio Fault Zone	F Hess Escarpment	f Bocono Fault
--6a Colima Rift	--11e Corsair Fault Zone	G Beata Ridge	g Merida Andes
7 Infiernillo triple junction	12 Western Chortis Rifts	H North Managua Boundary Fault	h Altemira Fault
--7a Tepalcatepec Rift	13 Chiapas fold–and–thrust belt	I South Managua Boundary Fault	i Romeral fault zone
8 Oriental Front	14 Missing Mexican forearc	J North Panama Trench	j Carnegie Ridge
9 Mexican Ridges		K Cocos Ridge	k Grinjalve Fracture Zone(s)

evolving and deforming to accommodate the apparent instabilities and to estimate for how long this accommodation may have been taking place.

Specific tectonic features discussed in the text are illustrated in Fig. 2.1 relative to continental outlines, and then again, relative to elevation data (Figs. 2.2 and 2.3), magnetic data (Figs. 2.4 and 2.5), and free-air anomaly gravity data (Figs. 2.6 and 2.7). These geophysical layers provide surface and subsurface constraints on preserved geology. One of the key observations is that evidence for lineaments connecting the northern and southern Caribbean plate boundary zones to the Middle America Trench are absent in all of the geophysical data sets (Figs. 2.3, 2.5, and 2.7). Conventional Pacific models require such lineaments to have existed in the recent past (e.g., Pindell and Kennan 2009), so it is noteworthy that no evidence for such lineaments is directly available. One possibility is that recent sedimentation and/or magmatism have covered or otherwise overprinted the hypothesized faults, but the absence of evidence justifies consideration of In-situ and Pirate models of western Caribbean tectonics.

In this book, east-dipping subduction through time at the western North, Central, and South American margins collectively (Fig. 2.1), is generalized as Farallon-related subduction; possible distinctions between the Farallon Plate and its descendents, Cocos and Nazca Plates (Fig. 1.1), are only made explicit as needed. I refer to the South Mexico, Middle America, and Nazca Trenches to indicate trench segments of the Farallon-related subduction system that border North, Central, and South America, respectively (Figs. 1.1 and 1.2). This contrasts with some papers in which the trench segments off both North and Central America are considered part of the Middle America Trench (e.g., Kim et al. 2011).

The question of global mass conservation is also important in the Caribbean context because it is uncertain where extension conjugate to shortening at the eastern Caribbean Plate boundary has occurred. Extension conjugate to this shortening must be identified lest the implication be the Earth is shrinking about the Caribbean region. A roughly similar extension along the entire length of the Mid-Atlantic Ridge indicates that conjugate extension does not occur there (Fig. 1.1); sinistral and dextral shear along the northern and southern Caribbean Plate boundaries implies that extension conjugate to eastern Caribbean Plate boundary shortening lies to the west—located wherever the northern and southern boundary faults terminate. In In-situ models, extension and crustal thinning are inferred to have taken place directly in the western and central portions of the Caribbean Plate (James 2009, 2013). In Pirate models, extension and crustal thinning are inferred to have taken place in eastern Mexico/western Gulf of Mexico behind the departing Chortis Block lithosphere (Keppie 2012). Pacific models of western Caribbean tectonics infer a continuation of the northern and southern boundary faults across the western Caribbean Plate corners (e.g., Fig. 1.3; Mann 2007; Pindell and Kennan 2009), but do not address where the extension conjugate to eastern Caribbean shortening actually occurs.

Similarly, opening of the Gulf of Mexico needs to be considered in the context of Caribbean tectonics (Pindell and Dewey 1982; Dickinson 2009). Some authors have suggested that a southward and counter-clockwise rift opened the Gulf of Mexico in the Jurassic, but only after proto-Caribbean rifting started between the Yucatan Block and northern South America (e.g., Pindell and Kennan 2009; Dickinson 2009 and references therein). If so, this would mean the shortening conjugate to Gulf of Mexico opening must be identified to the southeast in the Caribbean realm. This is because rift opening of the proto-Caribbean seaway between North and South America constrains the total space that can be created between North and South America, and the subsequent opening of the Gulf of Mexico can not imply space created in excess of this. The alternative possibility is that opening of the Gulf of Mexico predates rifting in the proto-Caribbean which only started following a ridge jump from the Gulf of Mexico to the proto-Caribbean seaway (Bird et al. 2005). However, no evidence for such a ridge jump has been confirmed from

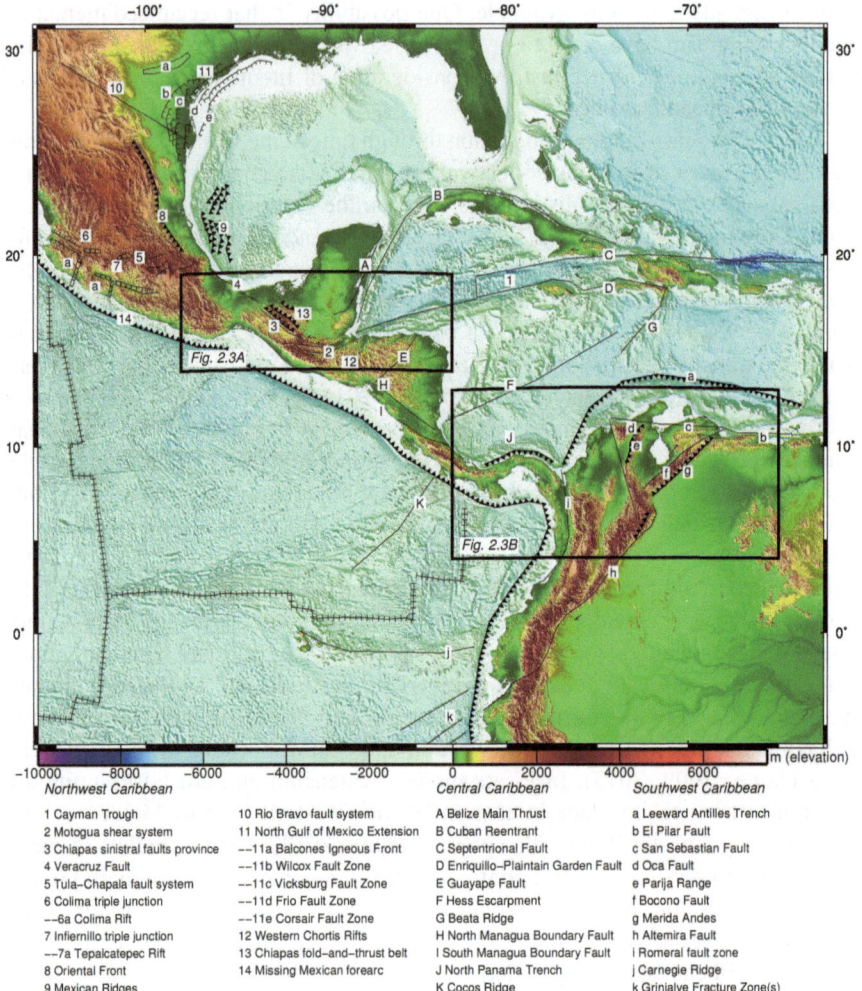

Fig. 2.2 Tectonic features in the greater western Caribbean region underlain by elevation data. (Elevation data from Farr et al. 2007)

modern accounts of sea-floor spreading in the mid-Atlantic (Labails et al. 2010). Neither Pacific nor In-situ models of Caribbean tectonics link the evolution of the Gulf of Mexico with the evolution of the western Caribbean region (Pindell and Kennan 2009; James 2009). In contrast, extension in the Gulf of Mexico may have been balanced by shortening at the eastern Caribbean Plate boundary zone in the Pirate model (Fig. 1.3c) to the extent that capture of the Chortis Block during lateral intrusion drove complementary rotation of the Yucatan Block to the east as a consequence of coupling across their common boundary.

Opening of the Gulf of Mexico can only be linked to western Caribbean tectonics by Pirate model tectonics, if gulf opening was synchronous with post-mid-Cretaceous initiation of the Lesser Antilles and Middle America Trenches (Macdonald et al. 2000; Li et al. 2008). Such synchronicity is inconsistent with standard models, which restrict opening of the Gulf of Mexico to the Jurassic period (Pindell 1985; Bird et al. 2005; Dickinson 2009; James 2013). The Pirate model is, thus, challenged by the interpretation of Jurassic-only opening in the Gulf of Mexico. This challenge must be taken up in two stages, but can build upon the interpretations of younger rifting set forth by Reed (1995), for example. First, evidence for the timing and kinematics of extension and rifting in the western part of the Gulf of Mexico, i.e. directly behind the hypothesized capture of the Chortis Block, must be evaluated. Second, evidence for the timing and kinematics of extension and rifting in the main part of the Gulf of Mexico must be evaluated as well. In this chapter, the first part of this challenge is taken up in a preliminary fashion, whereas the second part is deferred to a later date. The primary emphasis of this chapter is to demonstrate how microplate capture processes may reconcile the tectonic record preserved near the western Caribbean Plate corners.

The question of triple point stability is the key constraint on the evolution of the western Caribbean corner regions. Pacific and In-situ models require classic stability, as per McKenzie and Morgan (1969), to have been the case at these corner regions in the recent and prolonged past (see Keppie 2013 for a review). In contrast, the Pirate model raises the possibility that lateral intrusion processes may have governed the evolution of the western Caribbean region for much of the Cenozoic at least. To my knowledge, lateral intrusion or microplate capture processes have never been recognized explicitly in terrestrial tectonics (e.g. Jones et al. 1997), at least until the hypothesis of Pirate tectonics was introduced for the northwest Caribbean corner zone (Keppie and Keppie 2012). Consequently, the rotation of microplates around vertical axes at broadly convergent margins have been almost exclusively explained in terms of lateral extrusion or microplate escape processes (e.g. Wallace et al. 2010). Such lateral extrusion explanations have been proposed in many studies of the western Caribbean corner regions, for example, (e.g. Ego and Ansan 2002; Suter et al. 2008; Andreani et al. 2008a, b; La Femina et al. 2009; Guzman-Speziale 2009). And, lateral extrusion of the Chortis Block onto and over the northwest Caribbean Plate is implied by the conventional Pacific model (e.g. Mann 2007; Pindell and Dewey 2009) as made clear by the converging flow lines illustrated in Fig. 1.3a.

However, given the present continuity of the Middle America Trench (Morgan et al. 2008; Authemayou et al. 2011), lateral extrusion processes are broadly restricted to the time prior to the formation of the unstable triple junction configurations depicted in Fig. 1.4. In contrast, lateral intrusion processes are possible principally in the time after the underlying instabilities formed in the western Caribbean region. Thus, the timing of instability formation is a key constraint to be determined. It is noteworthy that a global review identified microplate rotations about vertical axes in the western Caribbean Plate corner regions as cases poorly explained by lateral extrusion (Wallace et al. 2010). Paleomagnetic data indicates systematic counter-clockwise and clockwise rotations during the Cenozoic for microplates near the northwest and southwest Caribbean Plate corners, respectively (Gose 1985).

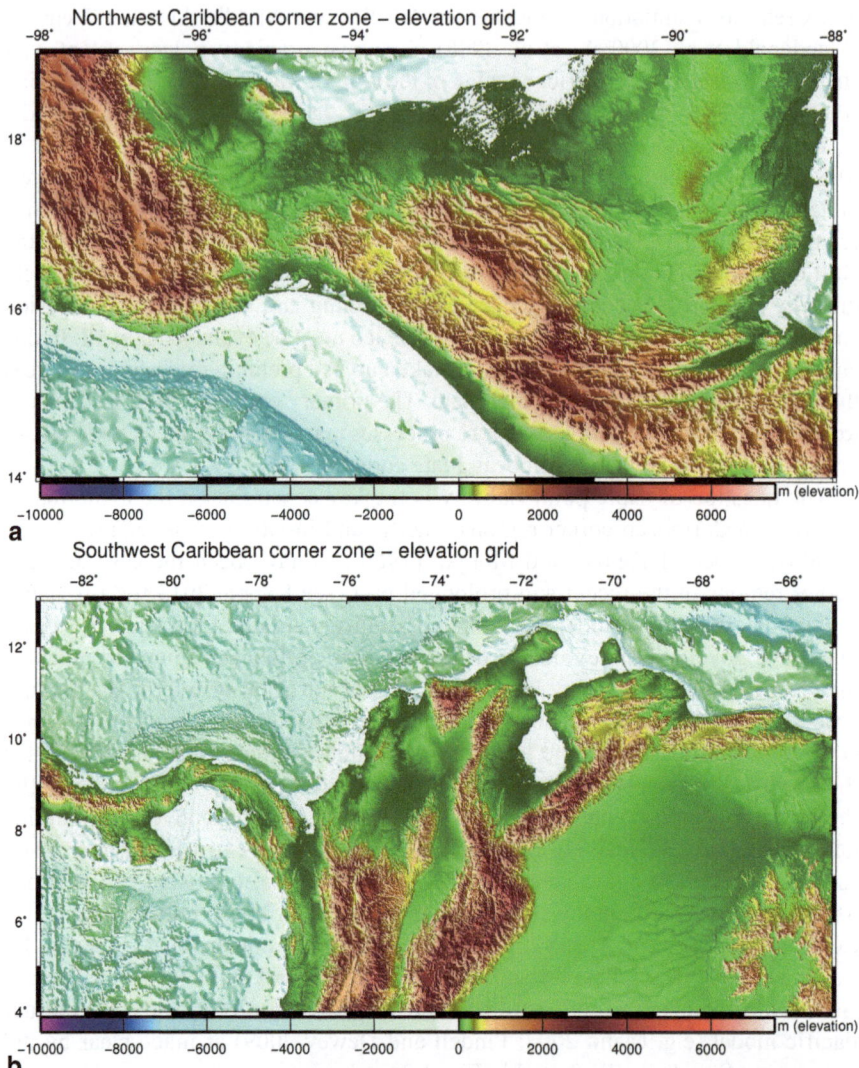

Fig. 2.3 a Elevation data at the northwest Caribbean Plate corner. **b** Elevation data at the southwest Caribbean Plate corner. Topographic lineaments for the northern and southern Caribbean Plate boundaries may not continue directly west to the Middle America Trench. (Elevation data from Farr et al. 2007)

2.2   Theoretical Analysis of Western Caribbean Tectonics

As a first approximation that helps to simplify this discussion, a Euclidean reference frame is used to discuss deformation across the Caribbean region (Fig. 2.8). Chapter 3 explains the selection and use of the Euclidean reference frame in detail.

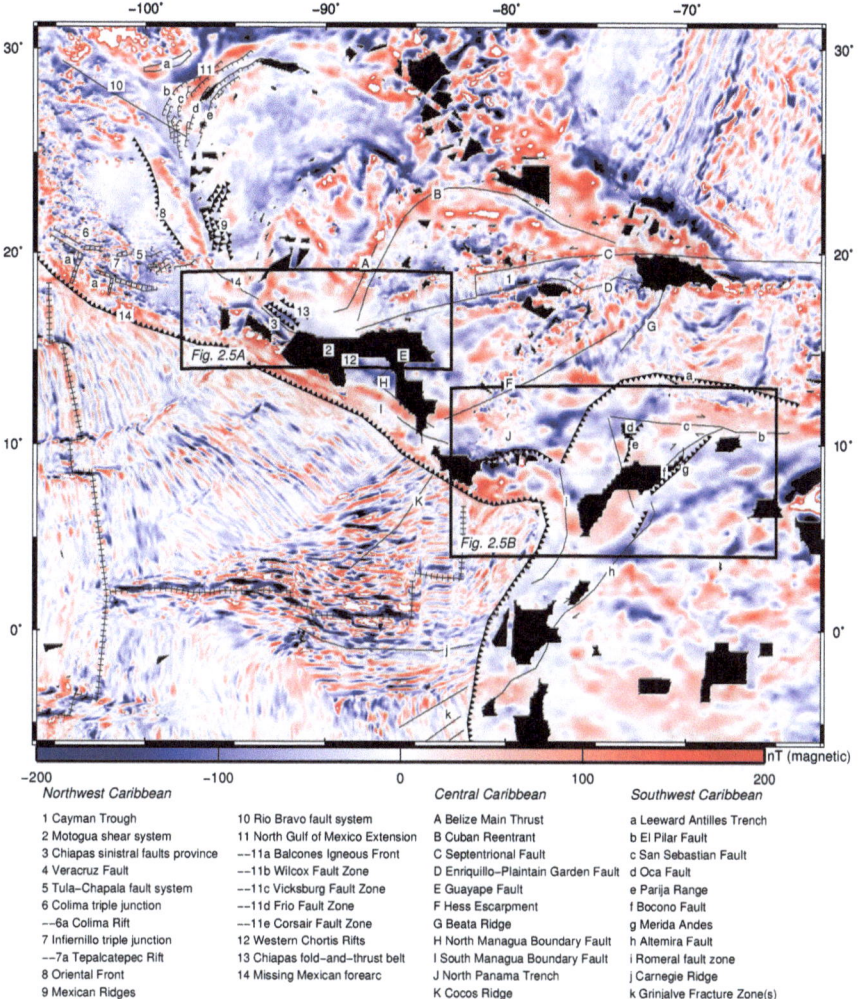

Fig. 2.4 Tectonic features in the greater western Caribbean region underlain by magnetic data. (Magnetic data from Maus et al. 2010)

Northwest Caribbean		Central Caribbean	Southwest Caribbean
1 Cayman Trough	10 Rio Bravo fault system	A Belize Main Thrust	a Leeward Antilles Trench
2 Motogua shear system	11 North Gulf of Mexico Extension	B Cuban Reentrant	b El Pilar Fault
3 Chiapas sinistral faults province	--11a Balcones Igneous Front	C Septentrional Fault	c San Sebastian Fault
4 Veracruz Fault	--11b Wilcox Fault Zone	D Enriquillo–Plaintain Garden Fault	d Oca Fault
5 Tula–Chapala fault system	--11c Vicksburg Fault Zone	E Guayape Fault	e Parija Range
6 Colima triple junction	--11d Frio Fault Zone	F Hess Escarpment	f Bocono Fault
--6a Colima Rift	--11e Corsair Fault Zone	G Beata Ridge	g Merida Andes
7 Infiernillo triple junction	12 Western Chortis Rifts	H North Managua Boundary Fault	h Altemira Fault
--7a Tepalcatepec Rift	13 Chiapas fold–and–thrust belt	I South Managua Boundary Fault	i Romeral fault zone
8 Oriental Front	14 Missing Mexican forearc	J North Panama Trench	j Carnegie Ridge
9 Mexican Ridges		K Cocos Ridge	k Grinjalve Fracture Zone(s)

Briefly, a Euclidean space tangent to the spherical Earth at a point can be used to evaluate stability at that point using Euclidean displacement or velocity change vectors (McKenzie and Morgan 1969). Further, if the Euclidean space is allowed to move along material flow lines, while remaining tangent to the spherical Earth at each successive point, then Euclidean displacement or velocity change vectors can be compared at different points along the common flow line. For general convenience, the Euclidean frame is chosen to be doubly tangent (Chap. 3). It is chosen so that it is tangent to both the spherical Earth and to a tectonic coordinate system in which it is also tangent to cones corresponding to small circle traces on the sur-

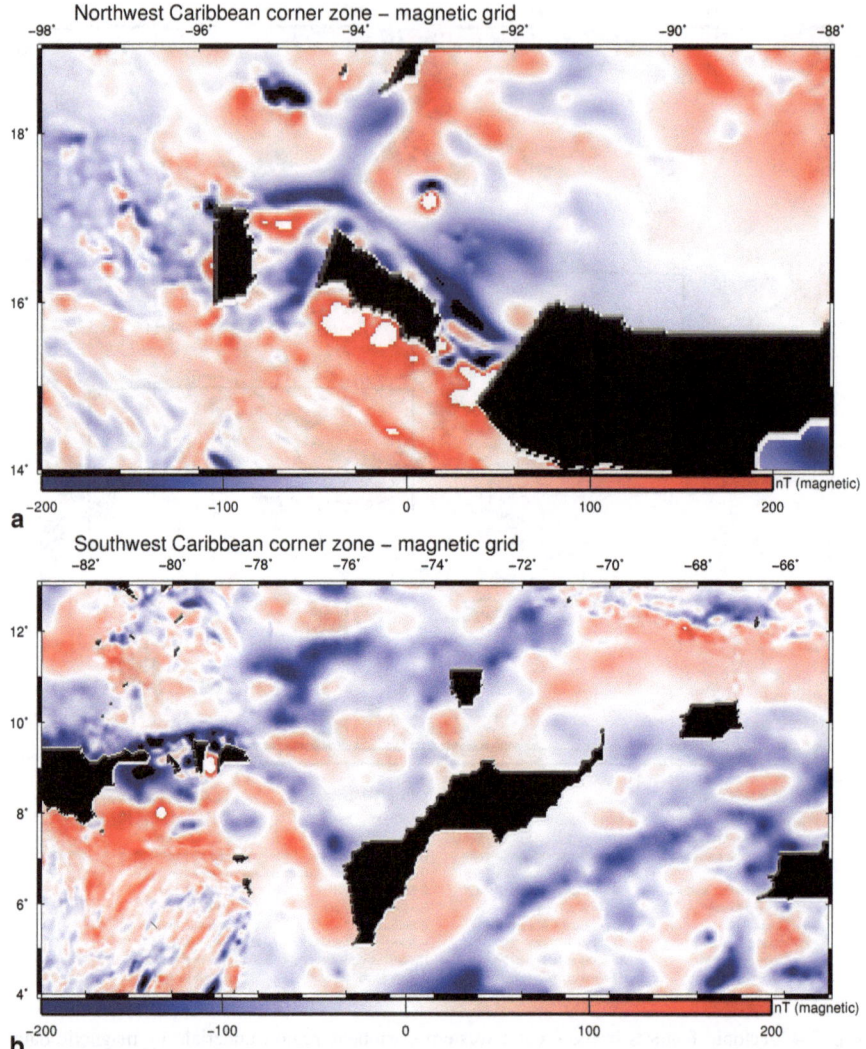

Fig. 2.5 a Magnetic data at the northwest Caribbean Plate corner. **b** Magnetic data at the southwest Caribbean Plate corner. Topographic lineaments for the northern and southern Caribbean Plate boundaries may not continue directly west to the Middle America Trench. (Magnetic data from Maus et al. 2010)

face of Earth that express the rigid-body motion of one plate relative to another (or between a tectonic plate and the mantle). For convenience, the tectonic reference frame given by the net relative motion between the North America and Caribbean Plates since ca. 47.9 Ma has been chosen here. This specific reference frame has the advantage in the Caribbean case in that small circle traces used to approximate the

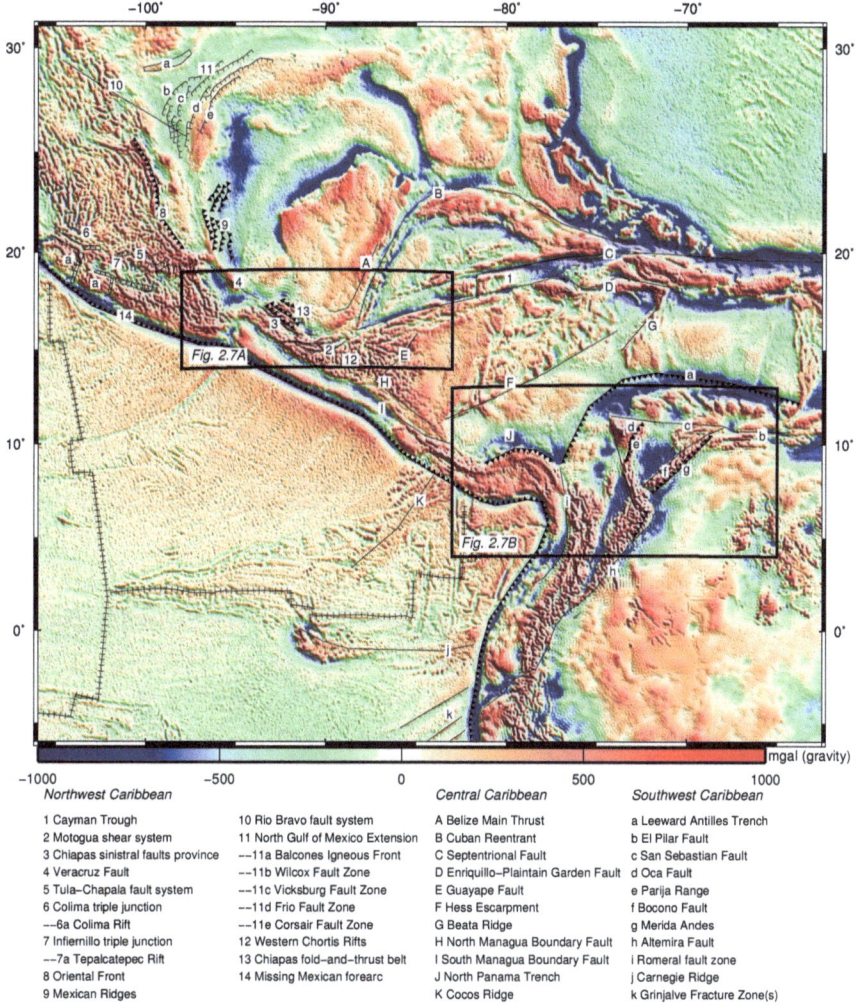

Fig. 2.6 Tectonic features in the greater western Caribbean region underlain by gravity data. (Gravity data from Sandwell and Smith 2009)

northern and southern Caribbean Plate boundaries appear straight when plotted in an oblique Mercator projection (Fig. 1.4c).

As discussed in Chap. 3, if the concept of a moving, doubly tangent reference frame is unfamiliar, it is adequate to view the Euclidean reference frame as an approximate tangent or secant projection of the Caribbean region as a whole (Fig. 2.8). This view does not introduce significant error for the scale of the discussion. To appreciate this, one can consider the formal errors introduced by the cylindrical Mercator projection, for example. In a Mercator projection, it is rigorous to note that the stretching errors associated with the projection are $1/\cos\theta$ for distance and

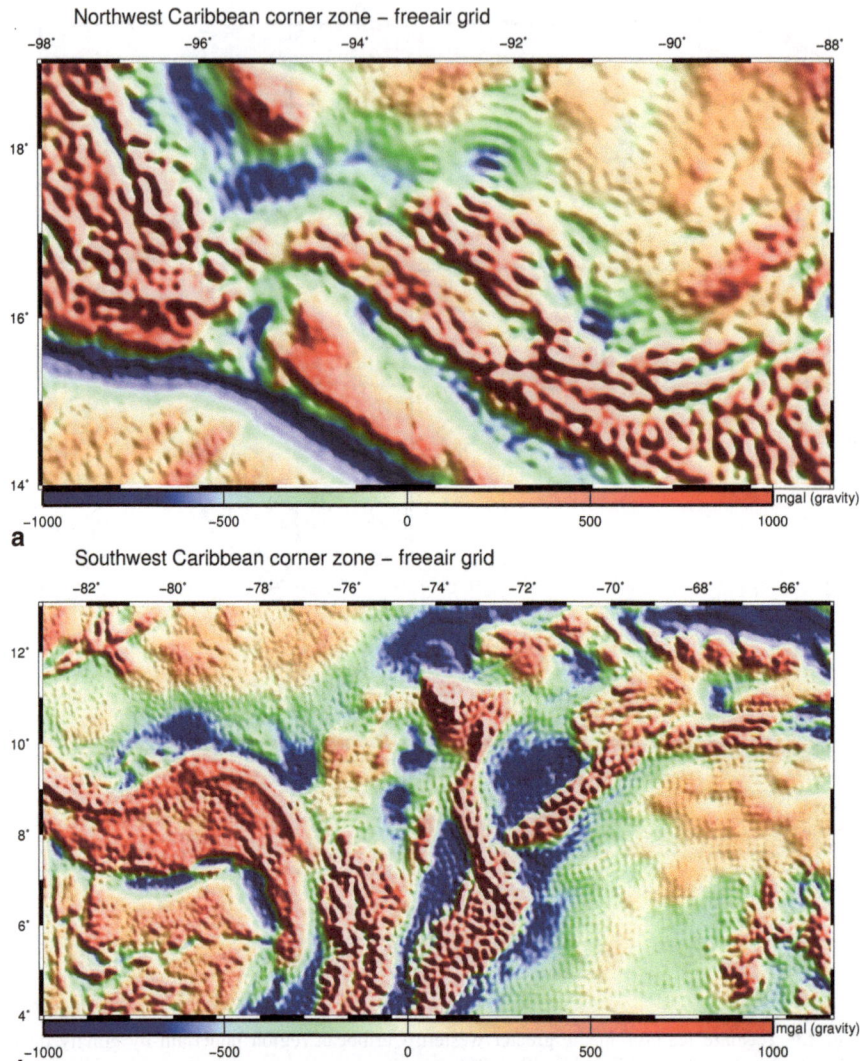

Fig. 2.7 a Gravity data at the northwest Caribbean Plate corner. **b** Gravity data at the southwest Caribbean Plate corner. Topographic lineaments for the northern and southern Caribbean Plate boundaries may not continue directly west to the Middle America Trench. (Gravity data from Sandwell and Smith 2009)

$1/\cos^2\theta$ for area, where θ is the projection latitude. So, for a Mercator projection equator that bisects the Caribbean region at approximately 15°N (at an appropriate prime meridian such as 80 W), distance errors of only ca. 1.5% and area errors of only ca. 3% would be implied 10° to the north or south (i.e. at ca. 5 and 25°N along the prime meridian). This range of latitude effectively spans the Caribbean

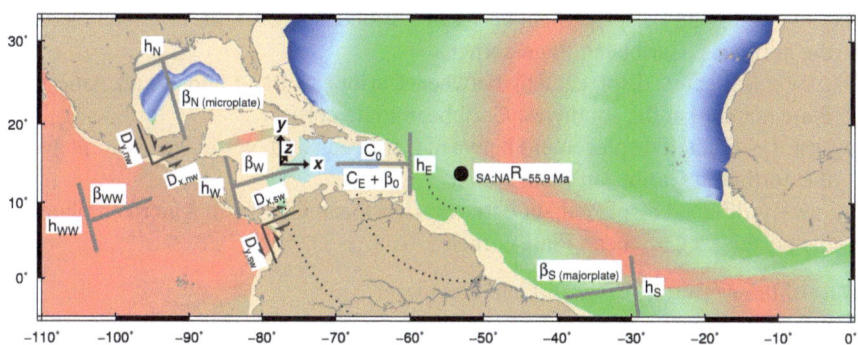

Fig. 2.8 Definition of (moving, doubly tangent, tectonic) Euclidean reference frame and variables for shortening (C), extension (β), distance (h), and displacement (D) used in evaluating western Caribbean Plate tectonics

region and corresponds to errors at 5 and 25°N of only 0.3 kmMa^{-1} relative to an approximate Caribbean-North America relative motion rate of ca. 20 kmMa^{-1}. This is negligible given typical errors in geological data. Evaluating tectonics in terms of a moving Euclidean reference frame also simplifies the analogy of the competing western Caribbean tectonic models with end-member models of mass conservation (Fig. 1.6).

With the above considerations in mind, the following terms are introduced and used in the text:

x direction ≈ projected longitude with −x ≈ west and +x ≈ east
y direction ≈ projected latitude with −y ≈ south and +y ≈ north
z direction ≈ depth with −z ≈ down and +z ≈ up
C=shortening, D=displacement, β=extension, h=distance
N=North, S=South, W=West, and E=East

At triple points:
   T=trench, F=fault, and R=ridge

At the eastern Caribbean Plate boundary:

C_0 km=x-directed component of convergence at the eastern Caribbean Plate subduction boundary
h_E km=the north–south length of this subduction zone hinge
β_0 km=x-directed component of divergence in the arc or back-arc region of the eastern Caribbean Plate boundary
$C_E=C_0-\beta_0$ km is the component of shortening *not* balanced by arc or back-arc extension in the eastern Caribbean region

At the western Caribbean Plate boundary:

Extension conjugate to eastern Caribbean shortening
β_W (*h_W) km=x-directed extension in the western Caribbean Plate region, and represents the back-arc region above the Middle America Trench

Dx, nw = x-directed displacement between Caribbean and North American plates at the northwest Caribbean Plate corner

Dx, sw = x-directed displacement between Caribbean Plate and South American Plate material at the southwest Caribbean Plate corner

Dy, nw = y-directed displacement between Caribbean and North American plates at the northwest Caribbean Plate corner

Dy, sw = y-directed displacement between Caribbean and North American plates at the northwest Caribbean Plate corner

West of the western Caribbean Plate boundary:

β_{WW} (*h_{WW}) km = x-directed extension located far to the west of the western Caribbean Plate boundary, possibly at the East Pacific Rise ridge zone

North or south of the western Caribbean Plate boundary:

β_N (*h_N) km = y-directed extension located to the north of the western Caribbean Plate boundary

β_S (*h_S) km = y-directed extension to the south of the western Caribbean Plate boundary

β_N and β_S could either be close by and intra-plate (such as β_N extension at the northern Gulf of Mexico), or be far away and at an existing plate boundary (such as β_S extension at the mid-Atlantic ridge segment in the South Atlantic

The above definitions allow the following predictive differences between the three end-member models of Caribbean tectonics to be articulated.

Global lithospheric volume is balanced via deformation of:

$C_E * h_E = \beta_{WW} *h_{WW}$ in the Pacific model
$C_E * h_E = \beta_W * h_W$ in the In-situ model
$C_E * h_E = \beta_N * h_N + \beta_S * h_S$ in the Pirate model

Relative displacements are balanced at the western Caribbean Plate corners via:

Dx, nw = Dx, sw = CE and Dy, nw = Dy, sw = 0 in the Pacific model
Dx, nw = Dx, sw = 0 and Dy, nw = Dy, sw = 0 in the In-situ model
Dx, nw = Dx, sw = 0 and Dy, nw + Dy, sw = CE in the Pirate model

2.2.1 Western Caribbean Plate Corners

2.2.1.1 Western Caribbean Plate Corners in the Pacific Model

According to the Pacific model, the northern and southern Caribbean Plate boundaries must continue west to the trench segments at the western Caribbean Plate boundary (Fig. 1.5a, b). As discussed above, extension conjugate to C_E shortening is β_{WW} and occurs to the west of the Farallon-related subduction system (Fig. 2.8). Two variations of the Pacific model correspond to whether the hinge trace length (h_{WW}) for β_{WW} extension equals or exceeds the hinge trace length h_E for C_E shortening.

If $h_{WW} >>> h_E$, then the western Caribbean Plate corners may have been stable TTT triple junctions (Fig. 1.5a). In this TTT-variation, the trailing part of the Caribbean Plate will have subducted beneath the western margins of the North and South American Plates, as the TTT triple junctions migrated relatively southeastwards and northeastwards, respectively, along the western North and South American margins. In conjunction with triple point migration, this scenario permits accretion of arc material to the Caribbean Plate, formed above the Farallon-related subduction segment, to the western margins of North and South America. Alternatively, this circumstance also permits removal of arc material by strike-slip detachment of forearc microplates from the western margins of North and South America.

If $h_{WW} \approx h_E$, then FFT triple junctions may have existed on the northern and southern Caribbean Plate boundaries (Fig. 1.5b). In this FFT-variation, Farallon-related lithosphere must have been torn (or bent) in order to have accommodated the eastward migration of Caribbean Plate; shear sense would flip from sinistral to dextral, or vice versa, at the FFT triple points (e.g. Pindell and Dewey 1982).

2.2.1.2 Western Caribbean Plate Corners in the In-Situ Model

In the In-situ model, the northern and southern Caribbean Plate boundaries must die out or stop at or prior to reaching the Farallon-related trenches (Fig. 1.5c, d). End-member variations of the In-situ model correspond to whether the northern and southern boundary faults terminate in the eastern or western Caribbean Plate regions. In the fixed variation (Fig. 1.5c), $\beta_W = 0$ and all shortening at the eastern Caribbean Plate boundary must be balanced by extensional processes at the eastern Caribbean Plate boundary. Candidate possibilities include active slab rollback and back-arc spreading above the Lesser Antilles subduction zone. In the widening variation (Fig. 1.5d), $\beta_W = C_E$ and intra-Caribbean rifting unrelated to eastern Caribbean Plate boundary processes must occur. Candidate possibilities include back-arc spreading in Central America above the Middle American trench or intra-Caribbean rifting.

2.2.1.3 Western Caribbean Plate Corners in the Pirate Model

In the Pirate model, the northern and southern Caribbean Plate boundaries must accommodate the arrival of material into the western Caribbean region from the north or south (Keppie 2012; Keppie and Keppie 2012) (Fig. 1.5e, f). Relative to the Caribbean Plate, material arriving from the north or south will rotate into the trailing edge of the Caribbean Plate in a counter-clockwise or clockwise fashion, respectively. Two variations are possible.

In the major plate variation (Fig. 1.5e), the northern and southern Caribbean Plate boundaries could accommodate north–south convergence between the Caribbean and North or South American Plates. Accommodated convergence could be asymmetric and correspond to subduction, or have some symmetry and correspond to collision. Excess extension at the Middle Atlantic Ridge could accommodate rel-

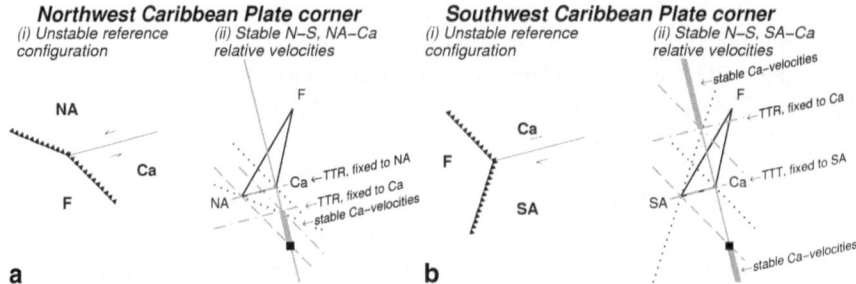

Fig. 2.9 A triple point stability analysis for: **a** the northwest Caribbean Plate corner, and **b** the southwest Caribbean Plate corner. Analyses assume the component of displacement between the Caribbean Plate and the North or South American Plates parallel to the northern or southern Caribbean Plate boundaries is known and fixed, whereas the component of displacement perpendicular to the northern or southern boundaries is unknown. *a.i* and *b.i* show reference unstable triple point configurations. *a.ii* and *b.ii* show the range of stable triple point configurations given the stated assumptions; they also show all permitted velocities of the Carribbean Plate relative to the North or South American Plates (*thin grey lines*) and the locus of these velocities that correspond to a stable triple junction configuration at the corresponding western Caribbean Plate corner (*thick grey lines*). End-member stable triple point configurations are indicated in which the triple point is fixed to the North or South American plates (*dashed grey lines*), or fixed to the Caribbean Plate (*dotted grey lines*). The special case for a possible Caribbean Plate velocity that allows stable configurations at both western Caribbean Plate corners without requiring differential motion between the North and South American Plates is also marked (*black square*) and illustrated in Fig. 1.5e

ative convergence of North or South American Plates into the western Caribbean region. Critically, however, this variation requires stable TTT triple points at the western Caribbean Plate corners. Triple point stability analysis of this hypothesis shows that north–south convergence is incompatible with the northwest Caribbean Plate corner, but compatible with the southwest Caribbean Plate corner where a stable TTT triple point can occur (Fig. 2.9a, b). Instead, rifting at the northern Caribbean Plate boundary and a corresponding TTR triple point can stabilize the northwest Caribbean Plate corner (Fig. 2.9a). Note that if the North and South American Plates move as a single American Plate, a single motion relative to the Caribbean Plate can stabilize both western Caribbean Plate corners. The stabilizing motion would involve asymmetric southward rifting and asymmetric southward subduction of the Caribbean Plate away from the North American Plate and under the South American Plate, respectively (*black squares* in Fig. 2.9).

In the microplate variation (Fig. 1.5f), synchronous extension within the North or South American Plates, and conjugate strike-slip shear oriented at an angle to the northern and southern Caribbean Plate boundaries, accommodates the escape of microplates from southern North America or northern South America, respectively. In this variation, the northern or southern Caribbean Plate boundaries curve into the North or South American Plates, respectively, where they become the eastern boundaries of microplate escape channels. These channels are bounded to the west by conjugate faults within the North or South American Plates with offsets equal to, but of opposite sense to, displacements across the northern or southern Caribbean

Plate boundary faults. In the microplate variation of the Pirate model with constant boundary conditions, unstable triple point configurations at the western Caribbean Plate corners are stabilized by microplates escaping from the southern or northern margins of North or South America in a diffuse pattern of inhomogeneous deformation. For the northwest Caribbean Plate corner, this means microplates could escape from the Gulf of Mexico. On the other hand, if boundary conditions were to change due to: (1) impingement of ridges or other indenting elements at the Acapulco or Nazca trenches, or (2) tears in Farallon-related lithosphere, then the northern or southern Caribbean Plate boundaries can connect to the North and South American margins, and microplates would be removed from their western margins.

2.2.2   Relative Plate Motions

The theoretical variations for the possible tectonic evolution of the western Caribbean (Fig. 1.5) require zero, one, two, or three rotation poles—as a minimum—to describe relative plate motions between the Caribbean Plate and the North and South American Plates. Thus, in the fixed, In-situ variation, the Caribbean Plate moves with the North and South American Plates and zero rotation poles are required. In the Pacific variations, the Caribbean Plate moves eastwards only with respect to the North and South American Plates, and only one rotation pole for Caribbean-American relative motion is required. In the major plate, Pirate variation, the Caribbean Plate moves eastwards with respect to the North and South American Plates, but the South American Plate also rotates into the western American region in clockwise fashion. This scenario requires two rotation poles, one for North American-Caribbean (NA-Ca) and another for South American-Caribbean (SA-Ca) relative plate motion. In the widening, In-situ variation, and in the microplate Pirate variation, plate material bends and turns a corner and three rotation poles are required, as discussed below. Here, a plate motion corner is considered to have two limb segments and a hinge zone using the terminology borrowed from simple fold geometry.

   A plate motion corner involves a z-to-x-directed change in relative plate motion in the In-situ model, and material enters and exits the western Caribbean region from depth and to the east, respectively. In general, for z-to-(x|y)-directed (or (x|y)-to-z-directed) plate motion corners, the corner hinge is parallel to the surface of Earth and the hinge trace is exposed as a diffuse line. In these cases, only relative plate motion in the surface limb segment is described by an Earth-centered pole of rotation and incorporated in global plate motion models. This is the case for lithospheric material exiting or entering classic divergent and convergent plate boundaries, respectively.

   On the other hand, a plate motion corner involves a y-to-x-directed change in relative plate motion in the Pirate model, and material enters the western Caribbean region from the north or south and exits to the east. In this case, for y-to-x-directed (or x-to-y-directed) plate motion corners, the corner hinge is radial to the surface of

Earth and the hinge trace is exposed as a diffuse point. In these cases, relative plate motion in both limbs and the hinge zone are described by Earth-centered poles of rotation and incorporated in global plate motion models. An Earth-radial plate motion corner is considered as a special case of a plate circuit (Cox and Hart 1986) that follows the path of changing plate motion. For example, rotation poles axial to the gentle curvature of the western Gulf of Mexico, the tight curvature of the Motogua-Polochic-Jocotan shear system near the northwest Caribbean Plate corner, and the gentle curvature of the northern Caribbean Plate boundary would constrain counter-clockwise rotation of material out of the Gulf of Mexico into the trailing edge of the Caribbean Plate. In general, at Earth radial plate corners, plate motion changes will be counter-clockwise or clockwise depending on whether the relative change is right-handed or left-handed, respectively.

Rigid-body rotation can only involve a single relative rotation. Therefore, where plate motion changes across a plate motion corner, accommodating non-rigid deformation is implied. Non-rigid deformation may be localized to the transition zones between plate circuit steps. Alternatively, non-rigid deformation may correspond either to flexural–slip folding, tangential-longitudinal folding, or involve more sophisticated fluid deformation, if sufficiently slow strain rates or melting of plate material is involved (Turcotte and Schubert 2002). Nevertheless, to a first approximation, the amount of lithospheric material entering and exiting a plate motion corner should balance.

2.2.3 Western Caribbean Arc Deformation Style

Arc deformation styles may be positive, neutral, or negative depending on whether arc regions experience relative shortening, no bulk deformation, or extension, respectively, above the otherwise convergent subduction zone systems (Dewey and Burke 1980). The models of possible tectonic evolution predict different styles for the Central American magmatic arc (CAMA) at the western Caribbean Plate boundary (Pindell et al. 2006). Broadly, (1) the fixed In-situ variation, the FFT Pacific variation, and the major plate Pirate variation correspond to positive deformation styles in the CAMA, and (2) the widening In-situ variation, the TTT Pacific variation, and the microplate Pirate variation correspond to neutral-to-negative deformation styles in the CAMA. In the first cases, this is because convergence of Farallon-related lithosphere at the western American and western Caribbean Plate boundary is similar and the arc deformation style is positive (Pindell et al. 2006). In the second cases, convergence at the western Caribbean Plate boundary is less than for the western American Plate boundaries, which allows for neutral-to-negative back-arc regions in the western Caribbean.

In the TTT, Pacific variation, convergence at the Middle America Trench segment is less than convergence at the Acapulco and Nazca Trench segments because convergence off western North and South America is the sum of both Farallon-related and Caribbean-related subduction, whereas off western Central America it is only due to Farallon-related subduction (Fig. 1.5a). In both the widening In-situ and microplate Pirate variations, extension in the Central American arc accommodates arrival of material from depth or from the north and south, respectively (Fig. 1.5c, f). CAMA extension is by far the greatest in the widening In-situ model. The Pirate model also predicts strike-slip shear within the Central American arc: dextral if related to the northwest Caribbean Plate corner, and sinistral if related to the southwest Caribbean Plate corner (Fig. 1.5f). The strike-slip shear accommodates the migration of material into the western Caribbean region from the north or south, respectively. The combined effects of microplate Pirate tectonics at both western Caribbean Plate corners would delineate a Central American forearc sliver at the trailing edge of the Caribbean Plate, with CAMA extension greatest at a midpoint between the northwest and southwest corners along the Middle America Trench.

2.3 Review of Western Caribbean and Gulf of Mexico Tectonics

2.3.1 Shortening at the Eastern Caribbean Plate Boundary

Total shortening at the eastern Caribbean Plate boundary is typically inferred to be > 1,500 km (Pindell et al. 2006), which corresponds to the width of oceanic lithosphere that may have subducted at the Lesser Antilles Trench since subduction initiated there in the late Cretaceous (ca. 120–80 Ma: Bouysse 1988; Bouysse and Westercamp 1990; Macdonald et al. 2000). The shortening of > 1,500 km is similar whether one uses the width of conjugate magnetic anomalies off western Africa (Müller et al. 2008), the width of compatible magnetic anomalies off eastern North America (Müller et al. 2008), or the down-dip extent of probable slab profiles in seismic tomography beneath the eastern Caribbean Plate (Li et al. 2008) (Fig. 2.10a). Estimates based on magnetic anomalies off western Africa and eastern North America may not be robust, however, since most of the lithosphere inferred to be under the eastern Caribbean Plate relates to proto-Caribbean (rather than mid-Atlantic) seafloor spreading (James 2013).

Active back-arc extension above the eastern Caribbean Plate boundary is implied in most tectonic models until the late Cretaceous (Pindell and Kennan 2009 and references therein). Generally, none is inferred after the late Cretaceous, which is consistent with active slab rollback acting primarily during juvenile subduction (Gurnis and Hager 1988). Uncertainty regarding the origin of Caribbean lithosphere (Driscoll and Diebold 1998; Giunta and Beccaluva 2006; James 2009) means that estimates for the amount of back-arc extension in the eastern Caribbean region are poorly constrained. Estimates of shortening amounts at the eastern Caribbean Plate boundary since the late Cretaceous, can be inferred from sinistral shear displacements at the Cayman Trough at the northern Caribbean Plate boundary for the Cenozoic (Fig. 2.1). Long-term shortening rates appear to have been between 15–20 $kmMa^{-1}$ based on magnetic anomalies produced during mid-ocean ridge spreading within the Cayman Trough pull-apart (Rosencrantz and Sclater 1986; Rosencrantz et al. 1988; Leroy et al. 2000). Modern shortening rates appear to be ~20 $kmMa^{-1}$ based on global positioning system (GPS) data (Dixon et al. 1998; DeMets et al. 2007; DeMets et al. 2010). Net displacement between the Caribbean and North American Plates is typically inferred to be ~1,100 km over the last ca. 50 Ma (Burkart 1978; Valls Alvarez 2009; Morán-Zenteno et al. 2009). Thus, C_E shortening at the eastern Caribbean Plate boundary likely falls between the estimates of Cenozoic and total shortening: ~1,100 km to > 1,500 km, respectively (c.f. Keppie 2012).

2.3.2   Extension Conjugate to Eastern Caribbean Shortening

The In-situ model for the tectonic evolution of the western Caribbean region implies that C_E shortening at the eastern Caribbean Plate boundary may be balanced by β_W extension. B_W extension is x-directed (i.e. west–east) and occurs within the western Caribbean region (Fig. 2.8). No estimates of such extension appear to be reported for the southwest Caribbean Plate region. In the northwest Caribbean Plate region, estimates of west-east extension are typically < 200 km (Morgan et al. 2008). Together, the western Chortis rift province and Sula rift record ca. 85 km of west-east sense intra-plate extension in the northwest Caribbean Plate region between 10 and 0 Ma (Rogers and Mann 2007) (Figs. 2.1, 2.2, 2.3, 2.4, 2.5, 2.6 and 2.7). Additional, limited, west–east extension may possibly occur within the Central American arc (Morgan et al. 2008), or across north–south valleys in the Nicaraguan Rise (James 2006, 2009).

The Pacific model for the tectonic evolution of the western Caribbean region implies that C_E shortening at the eastern Caribbean Plate boundary may be balanced by β_{WW} extension to the west of the western Caribbean Plate boundary (Fig. 2.8). It is not clear that any β_{WW} extension exists because studies that propose Pacific-type models have not explicity identified such extension (e.g., Pindell and Kennan 2009; Ratschbacher et al. 2009). One logical candidate for this extension could be the East Pacific Rise spreading center (Fig. 2.1); however, extension at the East Pacific Rise may be fully balanced by shortening at circum-Pacific subduction zones, and so, may not be a major factor.

The Pirate model for the tectonic evolution of the southwestern Caribbean region implies that C_E shortening at the eastern Caribbean Plate boundary may be balanced by β_S extension that is clockwise and to the south of the western Caribbean region (Keppie 2012; Keppie and Keppie 2012) (Fig. 2.8 and 1.5c). The South American Plate has converged towards the North American Plate in a clockwise fashion since ca. 55.9 Ma (Müller et al. 2008) (Fig. 2.11c). This means that extension at the southern segment of the mid-Atlantic trench may have balanced some C_E shortening; β_S extension can be attributed to the mid-Atlantic ridge in this case because the NA-SA rotation pole since ca. 55.9 Ma lies to east of the Lesser Antilles Trench (Müller

Fig. 2.10 Representative sections
through recent P-wave tomography
of the Caribbean region (Li et al.
2008): **a** shows >1,500 km as the
possible extent for a North American
slab subducted at the Lesser Antilles
trench, **b** shows that the Farallon
slab may be detached west of
Columbia, and **c** shows >500 km as
the possible extent for a Caribbean
slab subducted at the Leeward
Antilles trench

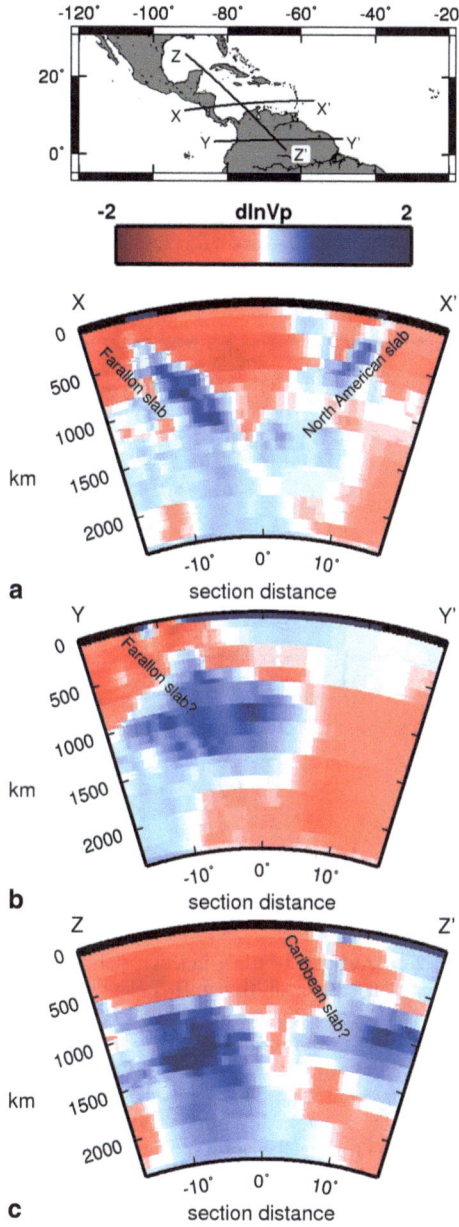

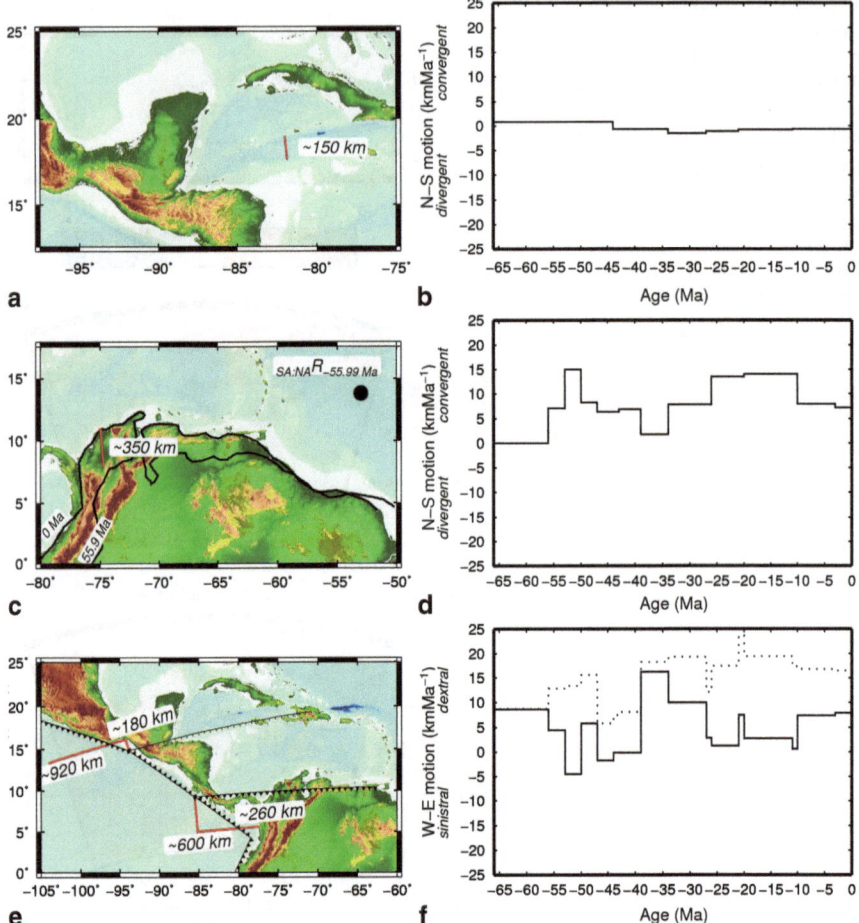

Fig. 2.11 Analysis of triple junction stability in the western Caribbean. **a** Present north–south extent of the Cayman Trough. **b** Relative convergence of Caribbean Plate towards the North American Plate at the point 82°W, 19°N, fixed to the North American Plate in a margin trending 75° from North. **c** Clockwise rotation of the South American Plate towards Nouth America since 55.9 Ma (Müller et al. 2008). **d** Relative convergence of South American Plate towards the North American Plate at the point 85.5°W, 9.25°N, fixed to the North American Plate in a margin trending 85° from North. **e** Net-balanced and residual components of strike-slip shear between the Caribbean Plate and the North and South American Plates at the northwest Caribbean Plate corners (see discussion in text). **f** Relative dextral shear of the Caribbean Plate relative to the South American Plate at the point 85.5°W, 9.25°N in a margin trending 85° from North (*dotted line*); dextral shear not balanced by South American clockwise convergence towards North America (*solid line*)

et al. 2008). Clockwise rotation of the South American Plate was accommodated by shortening at the Leeward Antilles trench at the southern Caribbean Plate boundary (Silver et al. 1975; Gorney et al. 2007). β_S extension available to balance C_E shortening is, thus, reduced by the amount of Leeward Antilles shortening. Clockwise convergence of the South American Plate into the western Caribbean region was ca. 350 km near the southwest Caribbean Plate corner, as constrained by relative motions between the North and South American Plates (Müller et al. 2008) (Fig. 2.11e).

The Pirate model for the tectonic evolution of the western Caribbean region implies that C_E shortening at the eastern Caribbean Plate boundary may be balanced by β_N extension that is counter-clockwise and to the northwest of the western Caribbean region (Figs. 1.5f and 2.9). North of the western Caribbean region, the Yucatan Block rotated southwards away from the southern USA and the Gulf of Mexico opened in a counter-clockwise fashion (Walper and Rowett 1972; Pindell 2010; Bird et al. 2005; Dickinson 2009). Opening of the Gulf of Mexico that predates subduction initiation at the Lesser Antilles Trench cannot be the β_N extension that balances C_E shortening. Some studies have inferred that all of the opening of the Gulf of Mexico occurred in the Jurassic, and therefore, predates subduction initiation at the Lesser Antilles trench (Pindell 2010; Burke 1988; Sawyer et al. 1991; Marton and Buffler 1994; Bird et al. 2005; Mickus et al. 2009; Pindell and Kennan 2009). This is consistent with low heat flow and deep ocean depths that have generally been inferred to correspond to Jurassic-aged oceanic lithosphere in the Deep Gulf of Mexico Basin (Stein and Stein 1992; Dickinson 2009). Stratigraphic columns throughout the Gulf of Mexico are generally believed to be floored by a single deposit of Callovian-aged (ca. 160 Ma) salt (Pindell 1985; Salvador 1987; Sawyer et al. 1991).

In contrast, other studies have inferred some late Cretaceous and Cenozoic opening of the Gulf of Mexico (Wilson 1993; Reed 1994, 1995; Wilson 2003; Fillon 2007; Rangin et al. 2008a, b). The Balcones complex is an igneous suite possibly associated with extensional tectonics that intruded into central Texas in the late Cretaceous (Spencer 1969; Griffin et al. 2010) (Fig. 2.1). The Corsair rift zone is a series of southward-younging listric normal faults identified in the northwest Gulf of Mexico platform that appears to have accommodated active tectonic extension in the late Cretaceous and Cenozoic and thinning of the lower crust (Rangin et al. 2008a, b). The southernmost and youngest fault in the Corsair rift zone also correlates with a heat flow anomaly possibly related to Cenozoic extension (Husson et al. 2008a, b). Cretaceous-to-Eocene unconformities and Eocene-Oligocene paleocanyons observed in the stratigraphy of the Tampico and Veracruz basins in the southwest Gulf of Mexico may also indicate Early Cenozoic rifting (Chapa 1985; Cantu-Chapa 1987). Back-stripping analysis indicates the basin floor in the western Gulf of Mexico is>2 km deeper than predicted by lithospheric age alone and records an early to mid-Cenozoic rifting event (Feng et al. 1994; Keppie and Keppie 2012). At least one author has proposed that stratigraphic columns throughout the Gulf of Mexico may contain salt with Cenozoic ages of deposition (Wilson 1993, 2003, 2004).

A key challenge in determining the age of extension in the Gulf of Mexico is filtering the signal of gravity tectonics (MacRae and Watkins 1992; Hossack 1994). The low relative viscosity and density of salt means that vertical and horizontal salt migrations may be entirely due to gravitational instabilities (Talbot 2004; Alzaga-Ruiz et al. 2009a, b), which could argue against Pirate tectonics, but in practice may only mean that deeper rifting signals are obscured (Reed 1994, 1995; Rangin et al. 2008a, b). Cenozoic gravity tectonics are widely reported in the Gulf of Mexico (MacRae and Watkins 1992; Hossack 1994; Talbot 2004; Alzaga-Ruiz et al. 2009a) and radial zones of higher heat flow are typically linked to salt diapir structures (Yu et al. 1992; Nagihara and Jones 2005).

2.3.3 Displacements at the Western Caribbean Plate Corners

2.3.3.1 Global Rotation Model

To calculate displacements at the western Caribbean Plate corners, relative plate motions are calculated from a global plate rotation model (Müller et al. 2008), in general. Specific to this chapter, the global model is supplemented by a North America–Caribbean rotation table derived from magnetic anomaly picks in the Cayman Trough (Leroy et al. 2000, NA-Ca pole table provided by the UTIG plates project courtesy of Rob Rogers and Sue Thompson, personal communication). Note that this North America–Caribbean rotation table only extends back to 49 Ma. In order to consider probable motion between 66 Ma and 49 Ma, I have adopted a back-of-the-envelope assumption that the oldest stage rotation and stage rotation rate in this table also applied prior to 49 Ma. Total Cenozoic displacement between North America and the Caribbean Plate calculated using this assumption is ~1,100 km, suggesting the assumption is reasonable (Burkart 1978; Morán-Zenteno et al. 2009).

2.3.3.2 X-Directed Displacements

X-directed displacements (Fig. 2.8) at the western Caribbean Plate corners may have been approximately equal to C_E shortening during the Cenozoic (Pindell and Kennan 2009; Ratschbacher et al. 2009). This would have been the case if sinistral and dextral shear propagated along the northern and southern Caribbean Plate boundaries, respectively, and reached the Middle America trench at the western Caribbean Plate corners (Fig. 1.5a, b). There is no evidence that the northern Caribbean Plate boundary reaches the Middle America trench at the northwest Caribbean Plate corner (Molnar and Sykes 1969; Guzmán-Speziale et al. 1989; Guzmán-Speziale and Meneses-Rocha 2000; Keppie and Morán-Zenteno 2005; Morgan et al. 2008; Guzmán-Speziale 2009; DeMets et al. 2010; Guzmán-Speziale 2010; Authemayou et al. 2011) (Figs. 2.2, 2.3, 2.4, 2.5, 2.6 and 2.7). The sinistral, northern Caribbean plate boundary continues westwards from the Cayman Trough into the Motagua shear zone, but prior to reaching the Middle America Trench, it appears to either stop (Morgan et al. 2008; DeMets et al. 2010) or curve to the northwest through the Chiapas fold belt and into northeastern Mexico (Andreani et al. 2008a, b; Guzmán-Speziale 2009, 2010). At the southwest Caribbean Plate corner, estimates of net dextral slip inferred from geological offsets on shear zones that may connect to the

Middle America trench are typically ca. < 300 km (Audemard and Audemard 2002; Audemard et al. 2005; Audemard 2009; James 2009).

2.3.3.3   Y-Directed Displacements

The Cayman Trough at the northern Caribbean Plate is now approximately 150–200 km wide in a north–south sense (Leroy et al. 2000) (Fig. 2.11a). If this width has been the same throughout the life of the Cayman Trough, the Cayman Trough may have formed as a pull-apart basin due to a north–south step in sinistral shearing at the northern Caribbean Plate boundary (Gough and Heirtzler 1969; Rogers and Mann 2007). The other possibility is that the Cayman Trough has progressively widened due to southward rifting of the Caribbean Plate relative to the North American Plate (Gough and Heirtzler 1969). Figure 2.11b plots the components of Caribbean convergence towards the point 82° W, 19° N on the North American Plate in a margin trending 75° from North. Negative values indicate divergence, which suggests the Caribbean Plate likely did rift southwards relative to the North American Plate. Rotations from the UTIG NA-Ca table indicate widening of the northern Caribbean Plate boundary has been ca. 1 km/Ma and relatively continuous since the mid-Cenozoic (Fig. 2.11b). Another possibility is that southward rifting was more episodic (Leroy et al. 2000). Magnetic anomaly axes in the Cayman Trough appear to have lengthened in the Eocene (> 49 Ma) and Oligocene (33–26 Ma) (Holcombe et al. 1973; Rosencrantz and Sclater 1986; Leroy et al. 2000). Additional southward widening at the northern Caribbean Plate boundary relative to the North American Plate may be inferred in the west–east elongate valleys at the northern margin of the Nicaragua Rise in the Caribbean Sea (Rogers and Mann 2007) (Fig. 2.2).

The Leeward Antilles Trench at the southern Caribbean Plate records the southward subduction of the Caribbean Plate under northern South America at this boundary (Silver et al. 1975; Gorney et al. 2007; Hippolyte and Mann 2010). Clockwise rotation of the South American Plate relative to the North American Plate since ca. 55.9 Ma accounts for ~350 km of southward subduction of the Caribbean Plate at the Leeward Antilles Trench (Müller et al. 2008) (Fig. 2.11c). To infer Caribbean–South American displacement, this amount should be added to the amount of southward rifting inferred for the Caribbean Plate away from North America. Minimum amounts of southward subduction of the Caribbean Plate beneath the South American plate could be ~500–550 km. Figure 2.11d plots the components of South American convergence towards the point 85.5° W, 9.25° N fixed to the North American Plate in a margin trending 85° from North. Convergence started at ca. 55.9 Ma and has continued since then. During this time, a lower convergence rate is notable during the late Eocene-Oligocene (45–25 Ma) and since the mid-Miocene (< 10 Ma). The 45–25 Ma slower convergence is approximately synchronous with the possible widening of the Cayman Trough (ca. 33–26 Ma) and the lower South American-North American clockwise convergence (ca. 45–25 Ma) suggests that southward subduction of the Caribbean Plate was ongoing at this time, in the context of a single North–South American Plate.

Seismic P-wave tomography indicates a slab beneath Columbia (Li et al. 2008) (Fig. 2.10b). Although the sub-Columbian slab has been attributed to Farallon-related lithosphere subducted at the Nazca trench off western South America (Pindell et al. 2006), this suggestion remains uncertain due to a possible gap in the Farallon slab under western Columbia (Fig. 2.10b). On the other hand, seismic P-wave tomography also indicates a sub-Columbian slab with down-dip extent > 500 km that may be connected to the Leeward Antilles Trench (Li et al. 2008) (Fig. 2.10c). Although the down-dip extent of a Leeward Antilles-derived slab in excess of ~500 km is incompatible with the major North American, Caribbean, and South American Plate motions (Fig. 2.11c), southward escape of microplate material across the northwest Caribbean Plate corner could have produced Leeward Antilles subduction in excess of ~500 km (Fig. 1.5f). Possible microplate tectonics in the western Caribbean region are reviewed separately below.

2.3.3.4   Residual Displacements at the Western Caribbean Plate Corners

Recall that southward migration of the Caribbean Plate can balance the eastward migration of the Caribbean Plate relative to the North and South American Plates at the western Caribbean Plate corners (Figs. 1.5e and 1.6). This means the total eastwards relative migration of the Caribbean Plate can be separated into a balanced part and a residual part at these corners (Fig. 2.11). The balanced part correlates with southward widening and southward shortening at the northern and southern Caribbean Plate boundaries, respectively. The residual part correlates with west–east slip that must appear as observed displacement across strike-slip faults.

At the northwest Caribbean Plate corner, Fig. 2.11e shows the result of a residual calculation for 1,100 and 150 km of eastwards and southwards relative Caribbean Plate motion, respectively, relative to North America since 66 Ma assuming a constant 40° angle between the Acapulco Trench and Cayman Trough. This calculation suggests ~180 km of eastward displacement is balanced by the possible southward displacement; only ~920 km of eastward displacement may appear as detectable fault slip. At the southwest Caribbean Plate corner, Fig. 2.11e shows the result of a residual calculation for ~860 and ~500 km of eastward and southward relative Caribbean Plate motion, respectively, relative to South America since 55.9 Ma, assuming a constant 40° angle between the Middle America Trench and the Leeward Antilles Trench. This calculation suggests ~600 km of eastward displacement may be balanced by the possible southward displacement; only ~260 km of eastward displacement may appear as detectable fault slip.

Figure 2.11f plots the detailed results of a more sophisticated residual calculation for the southwest Caribbean Plate corner. This calculation shows how much of the eastward migration of the Caribbean Plate relative to the South American Plate can be balanced solely by the northward migration of the South American Plate relative to the North American Plate at the southwest Caribbean Plate corner. This calculation indicates that there are three times when this convergence alone was insufficient to balance the eastward migration of the Caribbean Plate: (1) the Eocene for the period before 53 Ma, (2) the Oligocene for the period between 40 Ma and 26 Ma, and (3) the Miocene for times since 10 Ma.

2.3.4 Western Caribbean Microplates

The northwestern and southwestern corners of the Caribbean Plate have been fragmented into microplates and will be discussed below.

2.3.4.1 The Chortis, Chiapas, and Yucatan Blocks

Previous studies have introduced the Chortis and Yucatan (Maya) Blocks as Caribbean and North American microplates, respectively, presently separated by the Polochic-Motogua-Jocotan shear zone in Guatemala (Morán-Zenteno et al. 2009) (Figs. 1.2 and 2.1). In this chapter, I further sub-divide the previous Yucatan Block into northern, central-southern, and southernmost parts, i.e., the North Yucatan, the South Yucatan, and Chiapas Blocks, respectively, based on structural and topographic lineaments (Figs. 2.1, 2.2, 2.3, 2.4, 2.5, 2.6 and 2.7). Paleomagnetic data previously published for the Yucatan Block comes from the South Yucatan and Chiapas Blocks, as defined here (Gose 1985).

The Chortis Block may have been derived from the southern margin of Mexico, if the northern Caribbean Plate boundary connected directly west to the Middle America Trench throughout the Cenozoic (Ross and Scotese 1988; Pindell et al. 2006; Rogers et al. 2007; Silva-Romo 2008; Ratschbacher et al. 2009; Guerrero Garcia and Herrero-Bervera 2010). This corresponds to a Pacific model of western Caribbean region tectonic evolution and is the current leading hypothesis for the origin of the Chortis Block. The validity of this hypothesis depends critically on unidentified shear zone connections between the northern Caribbean Plate boundary and the Middle American Trench (Morgan et al. 2008; Guzmán-Speziale 2009, 2010). Stratigraphic correlations between the Chortis and Yucatan Blocks suggest proximity during the Cretaceous (James 2006, 2009; Keppie and Keppie 2012). Paleomagnetic data indicate the counter-clockwise rotation of the Chiapas, South Yucatan and Chortis Blocks since the Jurassic, and for the Chortis Block since the late Cretaceous (Gose 1985). Earthquake focal mechanisms suggest the Chortis Block is currently being extruded ESE away from the northwest Caribbean Plate corner (Guzmán-Speziale 2009, 2010).

The Pirate model of western Caribbean tectonics requires inversion of sinistral displacement at the northern Caribbean Plate boundary to dextral displacement on an oblique shear zone in southern North America (Fig. 1.5f). A candidate zone for dextral displacement in southern North America is the Sierra Madre Orientals, or Mexican Laramide, orogenic front in eastern Mexico/western Gulf of Mexico (Figs. 2.1 and 2.12). This front was active between ~70–40 Ma (Longoria and M. Suter 1990; Feng et al. 1994; Lang and Frerichs 1998). Therefore, it could have been synchronous with Late Cretaceous and Cenozoic extension in, and microplate escape from, the Gulf of Mexico. An estimate for the possible dextral displacement across the Sierra Madre Orientals front may be made by measuring the distance from its northernmost part, along its small circle trace, to the northwest corner of the Chiapas Block. The likely northernmost limit of the Sierra Madre Orientals is near Monteray, Mexico, where the orogenic front bends to the west and stikes across northern Mexico towards southern California (Figs. 2.1, 2.2, 2.3, 2.4, 2.5, 2.6 and 2.7). The distance from this bend to the northwest corner of the Chiapas Block is ~1,100 km (Fig. 2.12). Shortening in the Sierra Madre Orientals may be related to coupling at the Farallon-related subduction zone off western Mexico (Clark et al. 1982; Wark et al. 1990).

Shortening has occurred within the South Yucatan Block during the Miocene (Andreani et al. 2008a). Specifically, the Chiapas fold-and-thrust belt formed within the South Yucatan Block between ca. 11 and 9 Ma (Mandujano-Velazquez and Keppie 2009) (Fig. 2.1). Formation of the Chiapas fold-and-thrust belt has been linked variously to subduction of the Tehuantepec Ridge at the Acapulco Trench (Mandujano-Velazquez and Keppie 2009), or to collision between the North and South Yucatan Blocks (Kim et al. 2011). Evidence supporting a collision is the possible existence of a south-dipping Yucatan slab, imaged beneath the Isthmus of Tehuantepec (Kim et al. 2011).

2.3.4.2 The Transient Mexican Microplates

The Rio Bravo fault zone (northern Mexico), extends from near southern California to the Burgos basin in the western Gulf of Mexico, and was active between ca. 40 and 30 Ma (Flotte et al. 2008) (Fig. 2.1). This shear zone delineated a short-lived Mexico Block that migrated eastwards with sinistral displacement relative to the main North American Plate (Fig. 2.12c). Between ca. 25 and 17 Ma, the volcanic arc above the Acapulco trench in southern Mexico migrated rapidly landwards (Keppie et al. 2009; Morán-Zenteno et al. 2009) (Fig. 2.12a). Landward arc migration may have been diachronous, older in the west and younger in the east (Morán-Zenteno et al. 2009; Guerrero Garcia and Herrero-Bervera 2010). Rapid removal by subduction erosion of a ~250 km wide forearc block is inferred from the southern Mexican coast synchronous with the landward arc migration (Keppie and Morán-Zenteno 2005; Keppie et al. 2012).

Since the late Miocene, the Colima rift-rift-rift (RRR) triple junction may have been active in southwestern Mexico (Campos-Enriquez and Alatorre-Zamora 1998) (Fig. 1.18b). The eastern rift joining this triple junction is the Tula-Chapala zone of sinistral transtension (Campos-Enriquez and Alatorre-Zamora 1998; Andreani et al. 2008a) (Figs. 2.1 and 2.12). The Tula-Chapala zone of extension may connect in a complex fashion through the Veracruz fault in the Isthmus of Tehuantepec, then through the Chiapas strike-slip province, and finally to the Motogua-Polochic-Jocotan shear system in Guatemala (Dixon et al. 1998; Bandy et al. 2000; Andreani et al. 2008a). These faults delineate the South Mexico Block and demonstrate its

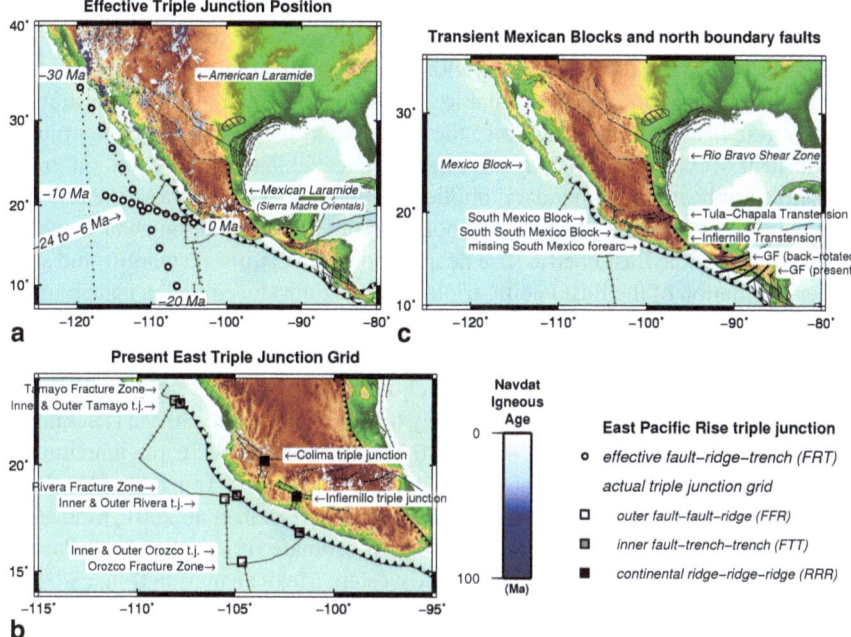

Fig. 2.12 Tectonics near the northwest Caribbean Plate corner (elevation data from Farr et al. 2007); volcanic age dates from NAVDAT. **a** Volcanic migration into the western North America appears to differ between the American and Mexican Laramide orogenies (English and Johnston 2004). Effective East Pacific Rise triple junction migration (open circle per Ma) estimated forward in time southeastwards from southern California since 30 Ma, and backwards in time northwestwards from southern Mexico since 0 Ma. Results suggest rapid southeastward migration since ca. 30 Ma before an unclear history between ca. 24 and 6 Ma. **b** Present-day East Pacific Rise triple-junction grid. A single fault-ridge-trench (*FRT*) triple junction between the East Pacific Rise, the Acapulco Trench, and the San Andreas fault system is actually a complex grid of triple junctions related to the Tamayo, Rivera, and Orozco East Pacific Rise transforms as indicated (Bandy et al. 2000). **c** Inferred transient Mexican Blocks formed progressively by communication between the East Pacific Rise triple junction and fault inversion at the northwest Caribbean Plate boundary. Clockwise back-rotation of the Guayape Fault (*GF*) in the Chortis Block indicates it may have been contiguous with the Acapulco Trench during the Oligocene, which together may have delineated the northern boundary of the missing South Mexico forearc

connection to the main Caribbean Plate (Andreani et al. 2008a). The South Mexico block may be currently rifting at the Inferniello RRR triple junction as well and the eastern branch (Inferniello zone) may be undergoing sinistral transtension (Bandy et al. 2000). If the Inferniello zone of sinistral transtension connects to the Polochic-Motogua-Jocotan shear zone, these faults delineate a South-South Mexico Block connected to the main Caribbean Plate (Fig. 2.12c).

In this book, the Mexico Block, the missing southern Mexico forearc, the South Mexico Block, and the South–South Mexico Block are called the transient Mexican Blocks (Fig. 2.12c). The formation of the transient Mexican Blocks appears to have started at ca. 40 Ma, and the northern margin of the transient Mexican Blocks has migrated generally southwards relative to North America since then. Consideration of triple junction stability means that shear zone connections that join the northern Caribbean Plate boundary to the Farallon-related subduction system off western North and Central America are unstable, if boundary conditions along these margins are roughly constant. Therefore, the observation that the northern margin of the transient Mexican Blocks may have been connected in a complex fashion to the northern Caribbean Plate boundary, implies that the points where these shear zones meet the western Mexican margins are points of boundary condition change.

The arrival of the East Pacific Rise near the southern California margin, and subsequent formation of the East Pacific triple junction, introduced such a plate boundary condition change along the Acapulco trench (McKenzie and Morgan 1969). The East Pacific Rise approached the southern California coast in the late Eocene/early Oligocene (Müller et al. 2008). Final arrival of the East Pacific Rise and formation of the East Pacific Rise triple junction likely occurred around ca. 30 Ma (Dickinson and Snyder 1979b). Between ca. 30 and 20 Ma, the East Pacific triple junction effectively migrated rapidly to the southeast, along the Acapulco Trench, before slowing when it neared the southern tip of Baja California (Oskin et al. 2001; Marsaglia et al. 2006) (Fig. 2.12a). Figure 2.12a plots an estimate of effective East Pacific Rise triple junction migration SSE down the western Mexican margin from ca. 30 to 20 Ma and ESE along the southern Mexican margin from ca. 10 to 0 Ma (open circle per Ma). Triple junction position through time is estimated as the moving intersection point of an East Pacific Rise great circle lineament with a North American margin great circle lineament assuming symmetric spreading between the Farallon-Cocos and Pacific plates. These estimates suggest an effective position for the East Pacific Rise triple junction off the southwest margin of Baja California between ca. 24 and 6 Ma. This age range indicates a correspondence between the probable location of the East Pacific Rise triple junction and where the northern boundary of: (1) the missing southern Mexico forearc (Keppie et al. 2012), and (2) the South Mexico Block probably was connected to the Acapulco Trench. Actual migration of the East Pacific Rise likely occurred in a stepwise fashion as progressive East Pacific Rise ridge segments and transform faults were subducted (Dickinson and Snyder 1979a; Bandy et al. 2000). Also, the present effective East Pacific Rise triple junction is actually comprised of a grid of nine local triple junctions related to the subduction of the Tamayo, Rivera, and Orozco transforms, indicating that triple junction migration is a complex process in detail (Fig. 2.12b).

2.3.4.3   Microplates Accreted to Northwest South America: Northern Andes, Maracaibo, and Bonaire Blocks

Geological studies of the northwest margin of South America describe two volcanic arc terranes that appear to have been accreted to this margin during the late Cretaceous and/or early Cenozoic (Kerr and Tarney 2005; Pindell et al. 2005; Kennan and Pindell 2009; Vallejo et al. 2009).

Since ca. 10–5 Ma, southeastward subduction of the Caribbean Plate at the Leeward Antilles trench has decreased, while dextral transpression on the Altemira-

Bocono-Merida Andes-El Pilar shear system has increased (Trenkamp et al. 2002; Audemard et al. 2005; Müller et al. 2008; Pindell and Kennan 2009) (Fig. 1.5f and 2.1). North of the Altemira-Bocono-Merida Andes-El Pilar shear system, the North Andean, Maracaibo, and Bonaire blocks have started to escape to the northeast and in a clockwise fashion relative to the main South American Plate (Audemard et al. 2005; Montes et al. 2005; Backé et al. 2006; Audemard 2009; Montes et al. 2010).

Possible connection of the Altemira shear zone to the Nazca trench off western Ecuador requires a change in boundary conditions at the Nazca trench near this junction, due to considerations of triple junction stability (Fig. 1.5e). The Carnegie ridge is currently subducting at the Nazca trench near this junction and may correspond to increased coupling between the Nazca and South American plates (Gutscher et al. 1999; Suter et al. 2008; c.f. Michaud et al. 2009). Alternatively, the Grijalve transform faults are now subducting at the Nazca trench and are sub-parallel to the Altemira shear zone (Rea and Malfait 1974; Hall and Wood 1985) (Fig. 2.1). The Grijalve transform faults have no conjugate on the Pacific Plate and may represent places where shear displacements across the southern Caribbean Plate boundary are forming tears within the Nazca Plate.

2.3.4.4   The Central American Forearc Sliver, Panama and Choco Blocks

Deformation within the Central American arc delineates the Central American forearc sliver as a small microplate at the western margin of the Caribbean Plate (La Femina et al. 2002, 2009) (Figs. 1.2 and 2.13). Arc deformation is primarily neutral-to-negative (Pindell et al. 2006; Morgan et al. 2008) and dextral (La Femina et al. 2002), except for the Cordillera de Talamanca in Costa Rica, which is positive (La Femina et al. 2009) (Fig. 2.13c). Previous studies attribute shortening in Costa Rica with the subduction of the Cocos Ridge, and related dextral shear in Nicaragua to lateral extrusion (La Femina et al. 2009) (Fig. 2.13a–c). However, the Managua bookshelf zone behind the Cordillera de Talamanca may be a zone of dextral transtension (Fig. 2.13c) and dextral shear in Nicaragua appears to exceed the dextral component of Cocos-Caribbean relative plate motion (Guzmán-Speziale 2010) (Fig. 2.13d). One possibility is that subduction of the Cocos Ridge produces deformation local to the Cordillera de Talamanca, but this mechanism cannot explain dextral shear in the Central American arc (Ego and Ansan 2002). An alternate possibility is that the Chortis Block is being extruded towards the ESE away from the northwest Caribbean Plate corner (Guzmán-Speziale 2009, 2010).

The Panama Block represents the Central American forearc south of the Managua bookshelf zone (Camacho et al. 2010). The complex Managua bookshelf zone reconciles dextral transtension behind the Central American forearc sliver with dextral transpression at the northern Panama Block subduction zone (Camacho et al. 2010). The Choco Block represents the northwesternmost part of South America (Fig. 1.2). At the eastern Choco Block boundary, sinistral shear on the Romeral fault zone is most likely (Ego et al. 1995), which may have accommodated the clockwise rotation of the South American Plate relative to the North American Plate since ca. 55.9 Ma (Figs. 2.1 and 2.11). The Panama and Choco Blocks may be broadly contiguous

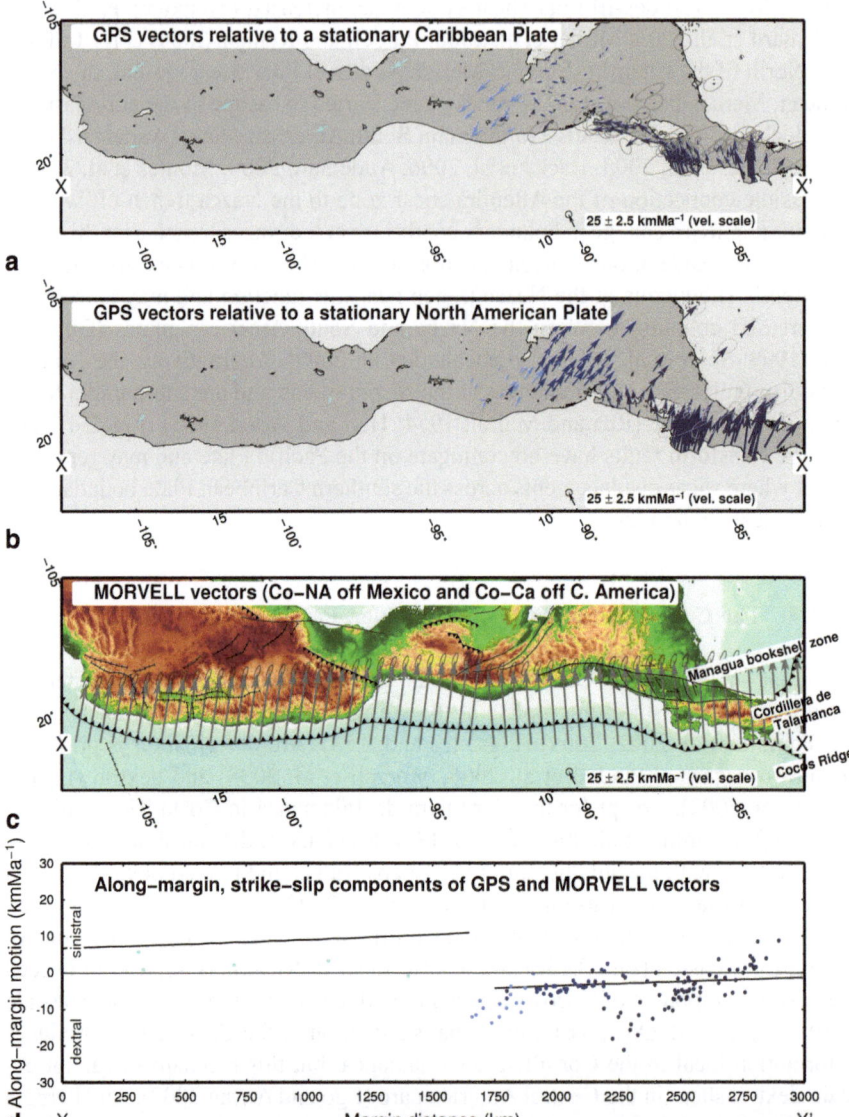

Fig. 2.13 GPS vectors for modern plate motions in the western Caribbean region. *Blue arrows* are continental GPS vectors relative to: **a** a stationary Caribbean Plate, and **b** a stationary North American Plate. GPS data from left to right, plotted in *light blue* to *dark blue* (from Marquez-Azua and DeMets 2009; Lyon-Caen et al. 2006; M. Rodriguez et al. 2009; La Femina et al. 2009). **c** *Grey arrows* show relative plate motion vectors from MORVELL for Cocos-North America (*Co-NA*) off southern Mexico and for Cocos-Caribbean (*Co-Ca*) off western Central America (DeMets et al. 2010). **d** Along the *margin*, strike-slip components of motion extracted from *Co-NA* and *Co-Ca* vectors (*black lines*) or projected from GPS vectors in **a** and **b** (*blue dots*) relative to a stationary *NA* in southern Mexico and relative to a stationary Caribbean Plate in Central America. This plot shows that dextral motion in the Central American forearc exceeds the dextral component of *Co-Ca* convergence

(Ego et al. 1995). Some studies suggest the Panama Block collided with the north-west margin of South America at ca. ~3 Ma (Kennan and Pindell 2009).

2.4 Discussion

Objective evaluation of regional tectonic models for areas such as the western Caribbean region, generally involves consistency with observed data, and predictive power (Kuhn 1977), which are briefly discussed below.

2.4.1 Consistency with Known Data

Wide acceptance of Pacific models implies most researchers believe shear zone segments that connect the northern and southern Caribbean Plate boundaries will be found (Rogers et al. 2007; Silva-Romo 2008; Kennan and Pindell 2009; Ratschbacher et al. 2009). Those who propose an in-situ origin appear to believe that unbalanced C_E shortening is overestimated and/or that β_W extension has been underestimated (Morgan et al. 2008; James 2009). The Pirate model balances C_E shortening with β_N extension, which is consistent with evidence for late Cretaceous and Cenozoic extension in the western Gulf of Mexico already (Rangin et al. 2008a; Husson et al. 2008a, b). Therefore, the Pirate model may be consistent with existing observations.

Each of the three end-member models are consistent with the evidence for large amounts of unbalanced C_E shortening since the late Cretaceous (Figs. 1.5 and 2.14): only the in-situ, fixed variation model is incompatible with this inference. Each of the three end-member models is consistent with the evidence for a neutral-to-negative Central American arc (Figs. 1.5 and 2.14) with the possible exceptions of the In-situ, fixed variation, the FFT-variation of Pacific, and the Pirate, major plate models.

2.4.2 Predictive Power

Both the Pacific and In-situ models fail to explain: (1) y-directed (N–S) plate motions, (2) y-to-x-directed (N/S–E) plate motions that are counter-clockwise in the northwest and clockwise in the southwest, and (3) tectonic phenomena to the north (+y direction) and south (−y direction) of the western Caribbean region that were synchronous with C_E shortening at the eastern Caribbean Plate boundary (Fig. 1.5a–d). As reviewed above, such plate motions and tectonic phenomena are part of the tectonic record of the western Caribbean region.

Previous analyses of western Caribbean tectonics (Fig. 2.15) leave major features of the tectonic record unexplained, including, for example: (1) counter-clockwise opening of the Gulf of Mexico, which appears to have no driving mechanism (Dick-

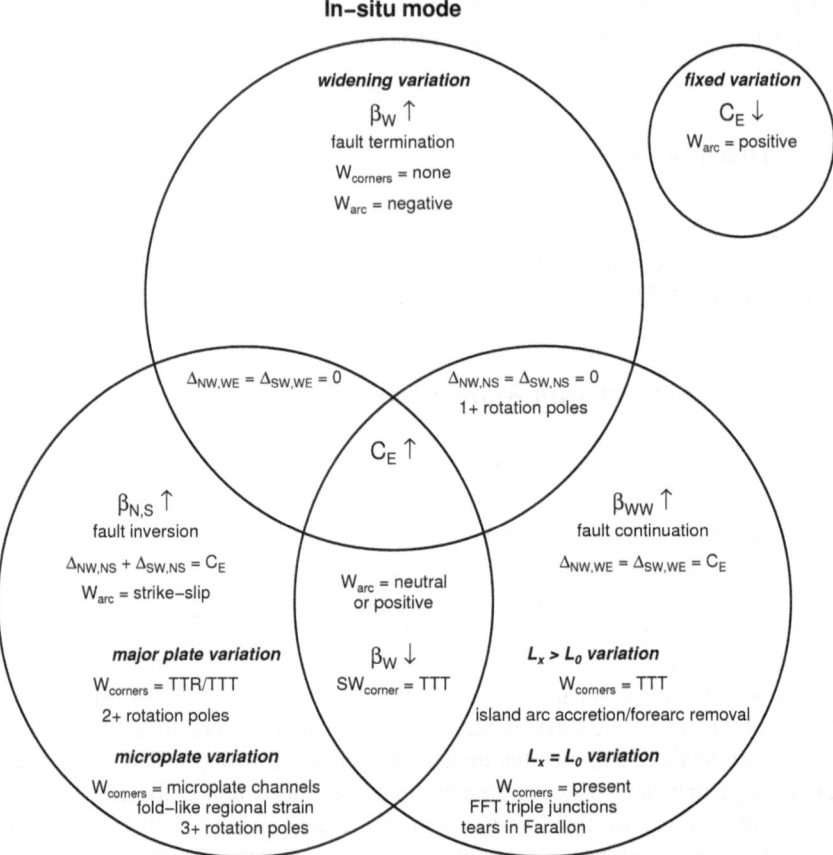

In–situ mode

widening variation
$\beta_W \uparrow$
fault termination
$W_{corners}$ = none
W_{arc} = negative

fixed variation
$C_E \downarrow$
W_{arc} = positive

$\Delta_{NW,WE} = \Delta_{SW,WE} = 0$

$\Delta_{NW,NS} = \Delta_{SW,NS} = 0$
1+ rotation poles

$C_E \uparrow$

$\beta_{N,S} \uparrow$
fault inversion
$\Delta_{NW,NS} + \Delta_{SW,NS} = C_E$
W_{arc} = strike–slip

$\beta_{WW} \uparrow$
fault continuation
$\Delta_{NW,WE} = \Delta_{SW,WE} = C_E$

major plate variation
$W_{corners}$ = TTR/TTT
2+ rotation poles

W_{arc} = neutral
or positive

$\beta_W \downarrow$
SW_{corner} = TTT

$L_x > L_0$ variation
$W_{corners}$ = TTT
island arc accretion/forearc removal

microplate variation
$W_{corners}$ = microplate channels
fold–like regional strain
3+ rotation poles

$L_x = L_0$ variation
$W_{corners}$ = present
FFT triple junctions
tears in Farallon

Pirate mode   **Pacific mode**

Fig. 2.14 Main variables and characteristics predicted for western Caribbean tectonics for the different models of local mass conservation plotted on a Venn diagram

inson 2009); (2) emplacement of the Balcones igneous complex (Griffin et al. 2010); (3) normal faulting and crustal thinning in the Corsair rift zone (Rangin et al. 2008a); (4) formation of the Sierra Madre Orientals (Fig. 2.12), unless it can be attributed to flat slab subduction under western Mexico (Feng et al. 1994), which may be unlikely (English et al. 2003; English and Johnston 2004); (5) counter-clockwise rotation of northwest Caribbean microplates (Gose 1985); (6) the concave outward curvature or bending of the northern Caribbean Plate boundary in the Motogua-Polochic-Jocotan shear zone (Fig. 2.1); (7) northward stepping of the northern Caribbean Plate boundary in the Motogua-Polochic-Jocotan shear zone (Ratschbacher et al. 2009); (8) dextral shear in the Central American arc that exceeds the dextral component of Cocos-Caribbean shortening (La Femina et al. 2009); (9) clockwise convergence of South America towards North America (Müller et al. 2008), and its stabilization of

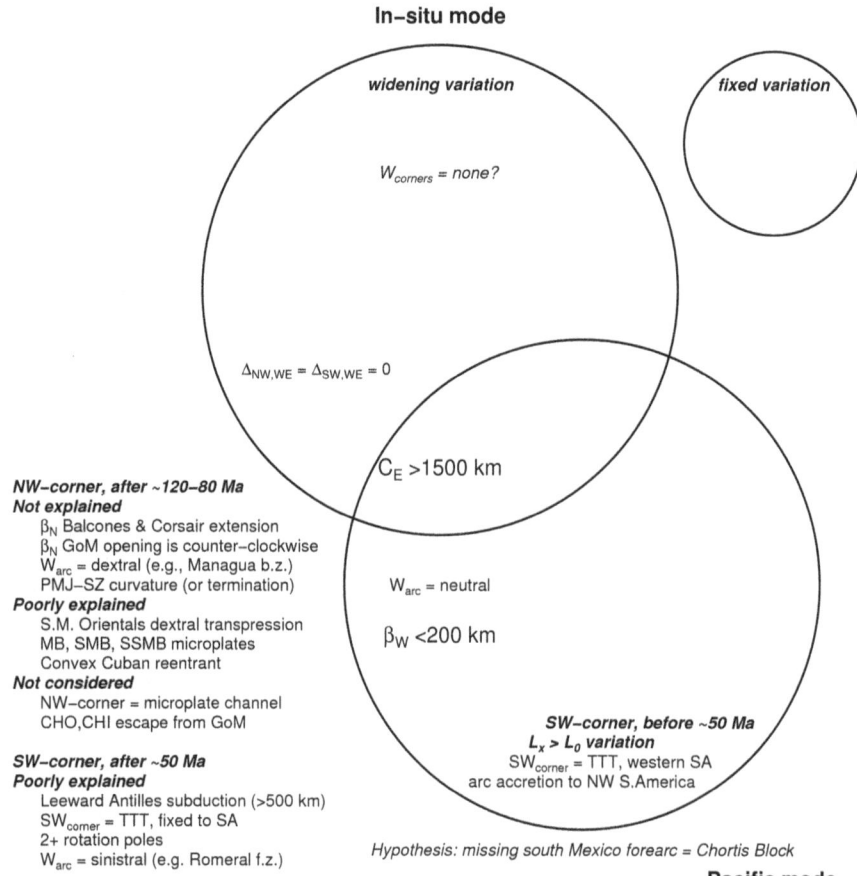

In–situ mode

widening variation

$W_{corners}$ = none?

fixed variation

$\Delta_{NW,WE}$ = $\Delta_{SW,WE}$ = 0

C_E >1500 km

NW–corner, after ~120–80 Ma
Not explained
  β_N Balcones & Corsair extension
  β_N GoM opening is counter–clockwise
  W_{arc} = dextral (e.g., Managua b.z.)
  PMJ–SZ curvature (or termination)
Poorly explained
  S.M. Orientals dextral transpression
  MB, SMB, SSMB microplates
  Convex Cuban reentrant
Not considered
  NW–corner = microplate channel
  CHO,CHI escape from GoM

SW–corner, after ~50 Ma
Poorly explained
  Leeward Antilles subduction (>500 km)
  SW_{corner} = TTT, fixed to SA
  2+ rotation poles
  W_{arc} = sinistral (e.g. Romeral f.z.)

W_{arc} = neutral

β_W <200 km

SW–corner, before ~50 Ma
$L_x > L_0$ variation
SW_{corner} = TTT, western SA
arc accretion to NW S.America

Hypothesis: missing south Mexico forearc = Chortis Block

Pacific mode

Fig. 2.15 Main variables and characteristics inferred for western Caribbean tectonics plotted on a Venn diagram for the Pacific and In-situ models. This diagram summarizes the outcome of previous the analyses which did not consider the Pirate model. Note that some features are either poorly, or not explained by the Pacific and In-situ models

the southwest Caribbean Plate corner (Fig. 2.11) has not been generally recognized; (10) it is not clear that sinistral displacement on the Romeral fault system at the eastern Choco Block boundary is well-explained (Ego et al. 1995; Suter et al. 2008); (11) southward rifting of the Caribbean Plate at the northern Caribbean Plate boundary (Leroy et al. 2000) and southward subduction of the Caribbean Plate at the southern Caribbean Plate boundary (Escalona and Mann 2011); and (12) the convex outward curvature of the Cuban Re-entrant must be explained by an appeal to counter-clockwise rollback of the Lesser Antilles trench (Pindell et al. 1988).

Also, how global surface mass conservation is satisfied in the Pacific and In-situ models has not been addressed. Crucially, shortening conjugate to the independent, counter-clockwise opening of the Gulf of Mexico, that post-dates rifting between the

North and South American Plates, has not been identified. For Pacific models, the location of β_{WW} extension west of the western Caribbean region, that is, conjugate and terminal to C_E shortening, remains unidentified as well (Pindell and Kennan 2009).

Despite their general lack of predictive power for features of western Caribbean tectonics, the Pacific and in-situ models appear to have secondary roles. The TTT-variation of the Pacific model is the only one that naturally predicts accretion of volcanic arc terranes to the northwest margin of South America during the late Cretaceous and early Cenozoic (Kerr et al. 1997; Kerr and Tarney 2005; Pindell et al. 2006). Active slab rollback and back-arc extension in juvenile subduction systems (Gurnis and Hager 1988) is consistent with the fixed variation of the in-situ model and the mid- to late-Cretaceous ages for eastern Caribbean Plate lithosphere (Macdonald et al. 2000). Since ca. 10 Ma, rifting in the western Chortis rifts and Sula rift (Rogers and Mann 2007) corresponds to the widening-variation of the In-situ model (James 2009).

On the other hand, the Pirate model accounts for most things unexplained by the other two models (Keppie 2012; Keppie and Keppie 2012)(Fig. 2.16). At the northwest Caribbean Plate corner, the microplate variation would account for synchronous: (1) counter-clockwise rotations of microplates during the Cretaceous and Cenozoic (Gose 1985) including opening of the Gulf of Mexico and escape of North American microplates (Pindell 1985; Dickinson 2009), (2) extension in the northwest Gulf of Mexico (Rangin et al. 2008a), and (3) dextral displacement in both eastern Mexico/western Gulf of Mexico (Keppie 2012) and the Central American arc (La Femina et al. 2009).

At the southwest Caribbean Plate corner, the major plate variation would account the synchroneity of clockwise rotation of the South American Plate into the western Caribbean region (Müller et al. 2008), greater mid-Atlantic rifting in the South Atlantic relative to the North Atlantic (Fig. 1.3), southward subduction of the Caribbean Plate at the Leeward Antilles trench (Hippolyte and Mann 2010; Escalona and Mann 2011), and sinistral displacement in the Romeral fault zone at the eastern Choco Block boundary (Ego et al. 1995).

The late Cretaceous and Cenozoic timing of the Pirate model likely corresponds to the mature stage of the Lesser Antilles subduction system at the eastern Caribbean Plate boundary (Fig. 2.17). During the mature stage, lack of active slab rollback and back-arc rifting implies that sinistral and dextral shear zones propagated towards the western Caribbean region along the northern and southern Caribbean Plate boundaries. The Pirate model predicts that these shear zones stop or curve in concave outward fashion prior to reaching the Farallon-related subduction system (Figs. 1.3 and 1.5), and the insufficient β_W extension conjugate to C_E shortening in the western Caribbean (Pindell et al. 2006).

Understanding the differences in how the Pirate model evolved at each western Caribbean Plate corner may be relatively straightforward as well. There are two main variations that require explanation. First, microplate rotations appear to be more important at the northwest Caribbean Plate corner, whereas major plate rotation appears to be more important at the southwest Caribbean Plate corner (Keppie 2012). Second, at the northwest Caribbean Plate corner, microplates (Yucatan,

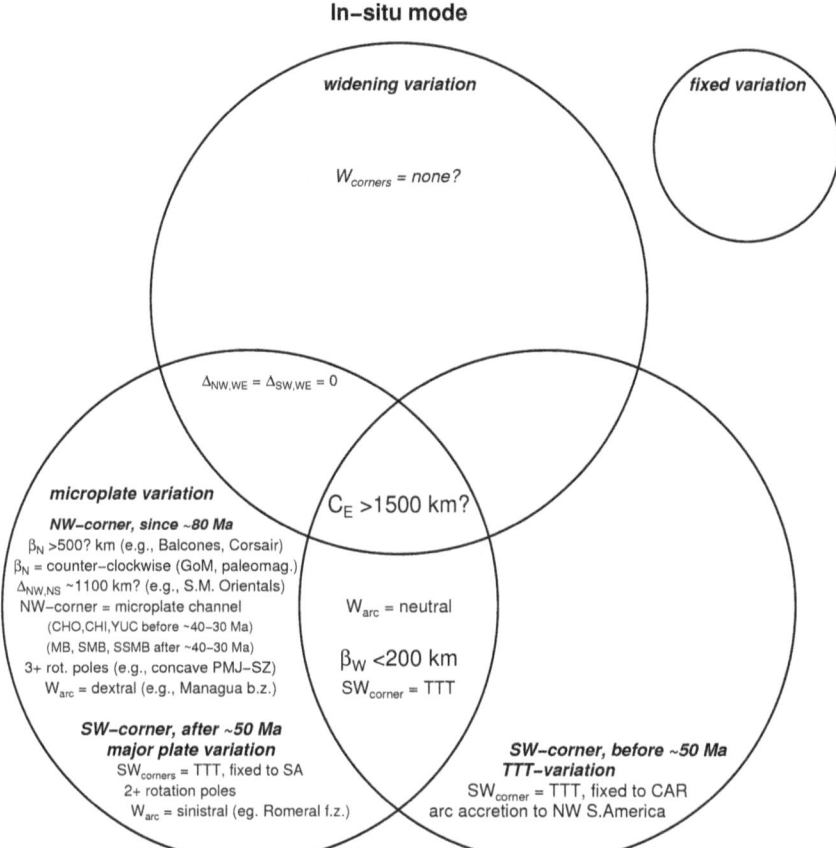

Fig. 2.16 Main variables and characteristics inferred for western Caribbean tectonics plotted on a Venn diagram for all three end-member models of local mass conservation. This approach corresponds to the analysis in this chapter. Note that tectonic features involving north–south senses of relative plate motion may correspond to the Pirate model

Chiapas and Chortis) escaping from the Gulf of Mexico by counter-clockwise rotation appears more likely prior to ca. 40 Ma (Pindell 1985; Keppie 2012; Keppie and Keppie 2012). On the other hand, microplates (South Mexico and missing Mexican forearc blocks) escaping from the southwestern North American Plate margin appears evident after ca. 40 Ma (Flotte et al. 2008; Andreani et al. 2008a; Keppie et al. 2009, 2012).

The first difference can be explained as follows. Given the y-to-x-directed nature of mass conservation in the Pirate model (Figs. 1.3 and 1.6), counter-clockwise rotation of the major North American Plate into the western Caribbean region at the northwest Caribbean Plate corner is impossible because triple junction stability would be violated

(Fig. 2.9a), which leaves the microplate variation of the Pirate model as the only viable model. In contrast, clockwise rotation of the major South American Plate into the western Caribbean region at the southwest Caribbean Plate corner is compatible with triple junction stability (Fig. 2.9b), i.e. mass balance and terminal β_S extension can employ the pre-existing mid-Atlantic ridge in the South Atlantic (Fig. 2.8).

If boundary conditions at the western North American Plate boundary are roughly equal everywhere, the second difference can be explained as well. Microplates must escape from the interior part of southern North America, accommodated by dextral shear in eastern Mexico/western Gulf of Mexico (Fig. 1.5f), i.e. the apparent triple junction instability at the northwest Caribbean Plate corner cannot be resolved by simple displacement to the northwest. However, given a suitable boundary condition change along the western North American Plate boundary, exterior microplate escape would be predicted if the sinistral northern Caribbean Plate boundary connects to the Farallon-related subduction system at the moment of boundary condition change. The arrival of the East Pacific Rise off western California in the late Eocene and subsequent formation and migration of the East Pacific Rise triple junction supplies a natural boundary condition change (McKenzie and Morgan 1969) (Fig. 2.12a) that would account for the apparent switching of microplate escape from the North American interior to the North American exterior margin.

Since the Oligocene, delineation and escape of microplates in southwestern North America would be transient because the location of the East Pacific Rise triple junction has migrated to the southeast, relative to the main North American Plate, since its formation (Fig. 2.12a). As a first approximation, the northern boundary of the exterior transient microplates should step southwards through time tracking the southward migration of the East Pacific Rise triple junction (Fig. 2.12a). However, the purported East Pacific Rise triple junction is not, in general, a single triple junction (Fig. 2.12b). Instead, at any given time, a grid of triple junctions may be present where multiple East Pacific Rise transforms are subducting and ocean lithosphere is coupling differently with the upper plate on each side of the transform faults (Fig. 2.12b). Thus, some out-of-sequence stepping of the northern boundary of the exterior microplates is expected (Fig. 2.12c).

The arrival of the Carnegie Ridge and the development of the Grijalve fracture zones off western Ecuador (Rea and Malfait 1974; Gutscher et al. 1999), may help to explain the emergence of exterior microplate tectonics at the southwest Caribbean Plate corner since ca. 10 Ma (Audemard 2009).

2.4.3   A New, Multi-Mode Model for Western Caribbean Tectonics

Based on the above considerations, the following multi-mode model for the tectonic evolution of the western Caribbean region is proposed (Fig. 2.17). The Caribbean Plate formed when subduction initiated at both the western and eastern Caribbean Plate boundaries in the mid- to late-Cretaceous. Since it is not robustly certain which is the latest subduction zone formed, it is not clear whether the early

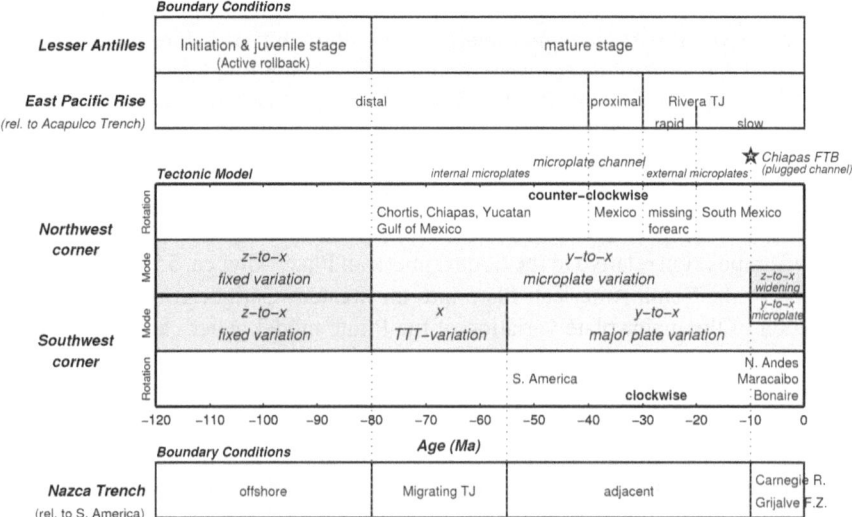

Fig. 2.17 Proposed model–time model for the tectonic evolution of the western Caribbean region. For each interval of time, the dominant model(s) and plate rotation(s), if any, are labeled. For each model transition, possibly-related regional boundary condition changes at the Lesser Antilles, Acapalco and Nazca trenches are also shown. Note that the Pacific, In-Situ, and Pirate models are labeled *x*-directed, *z-to-x*-directed, and *y-to-x*-directed, respectively, to emphasize the contrasting flow of lithospheric material through the western Caribbean region in each model

history of the Caribbean Plate corresponds best with the Pacific or In-situ modes of Caribbean tectonics. In the Pacific mode, eastern subduction is hypothesized to predate western subduction by some amount (in some models because west-dipping east-bounding Caribbean subduction started following a polarity reversal of earlier east-dipping west-bounding American subduction). In this case, eastward relative rollback of the eastern subduction zone may have been accommodated by ongoing sea-floor spreading in the Pacific Ocean realm and Farallon lithosphere may have entered the Caribbean region from the west. As discussed, this scenario best explains accretion of arc terranes to northwest South America in the late Cretaceous. Alternatively, the In-situ mode may have been important as well: Eastward rollback of the eastern trench may have been accommodated simply by back-arc extension in the immediate eastern Caribbean region while the eastern subduction zone was juvenile. In this case, the Caribbean region may have been divided into a western and eastern part analogous to the modern Scotia and South Sandwich Plates.

Regardless of the specific early history of the Caribbean Plate, however, once western Caribbean subduction was operating (i.e. probably after ca. 70 Ma) and as subduction at the eastern Caribbean Plate boundary became mature and a role for back-arc extension above it diminished (i.e. probably after ca. 70 Ma), then a consideration of the triple point evolutions discussed herein becomes critical. As interpreted here, the balance of evidence is best explained if a microplate channel developed across the northwest Caribbean Plate corner where interior North

American microplates escaped prior to ca. 40 Ma, and exterior microplates escaped after ca. 40 Ma. Progressive microplate capture through time is interpreted to have accommodated most of the eastward motion of the Caribbean Plate relative to North and South America since ca. 70 Ma. The change in microplate origin was due to the arrival of the East Pacific Rise and the formation of the East Pacific Rise triple junction off western North America.

At the southwest Caribbean Plate corner, the TTT-variation of the Pacific model operated until ca. 55.9 Ma, when the TTT triple junction arrived at approximately its modern position relative to the South American Plate. After ca. 55 Ma, clockwise rotation of the South American Plate into the western Caribbean region indicates operation of the major plate variation of the Pirate model. After ca. 10 Ma, clockwise rotation of microplates at the southwest Caribbean Plate corner has increased as clockwise rotation of the South American Plate has decreased.

The paleogeography implied by the proposed paleo-models is illustrated in Fig. 2.18, which provides a basis for visualizing tests of the main predictions. If many predictions are confirmed, then it will become useful to develop more detailed and quantitative control of the basic Pirate model and paleogeography presented in Fig. 2.18. A selection of notable predictions of the new multi-mode model are therefore introduced and briefly discussed below.

2.4.4   The Chortis Block Controversy

As discussed earlier in this chapter, the Chortis Block arrived in its current position accommodated by sinistral shear at the northern Caribbean Plate boundary. The possible late Cretaceous positions for the Chortis Block correspond broadly to the three end-member models of local mass conservation and their possible variations (Fig. 1.6).

The leading models for the origin of the Chortis Block invoke the TTT-variation of the Pacific model (Fig. 1.5a) and correlate the Chortis Block with the missing forearc removed from southern Mexico (e.g. Rogers et al. 2007; Silva-Romo 2008; Ratschbacher et al. 2009; Garcia and Herrero-Bervera 2010; Torres-de Leon et al. 2012 & Talavera-Mendoza et al. 2013) (Fig. 2.1). However, even if a shear zone segment connecting the Motogua-Polochic-Jocotan shear system to the Middle America trench is found, the Chortis Block is still unlikely to be the missing forearc from southern Mexico. This is because, in the Pacific model, the timing of forearc removal appears to be inconsistent with the timing of the landward migration of the southern Mexico volcanic arc during the late Oligocene/early Miocene (Keppie and Morán-Zenteno 2005; Keppie et al. 2009; Morán-Zenteno et al. 2009; Keppie et al. 2012) (Fig. 2.12a). Forearc removal may be too rapid and too young to correspond to the purported, relatively slow eastward migration of the Caribbean Plate relative to the North American Plate (Morán-Zenteno et al. 2009; Keppie et al. 2012).

In the multi-mode model, the Chortis Block is inferred to have escaped from the western Gulf of Mexico according to the microplate variation of Pirate tectonics

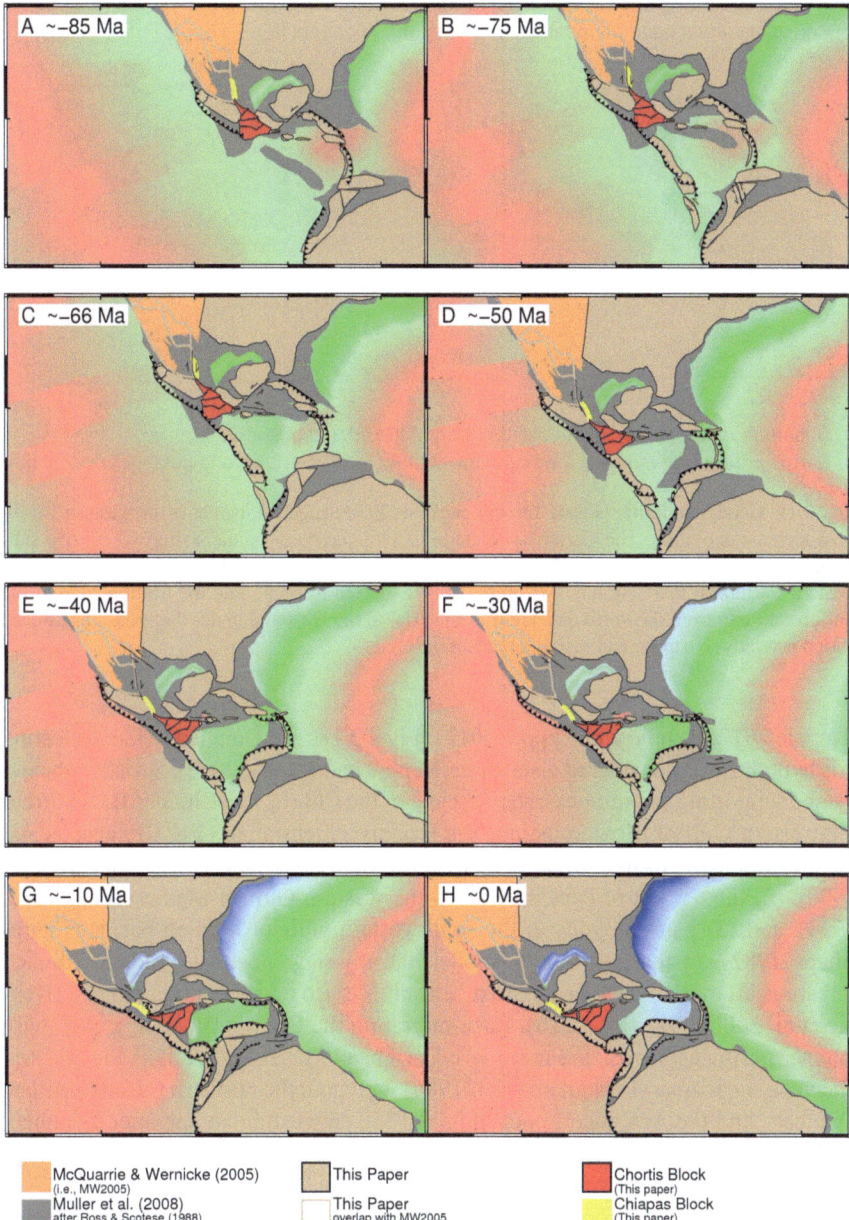

Fig. 2.18 A qualitative paleogeography for the western Caribbean region based on the proposed model–time model from Fig. 2.17, plotted in a moving mantle reference frame with global plate motions and ocean age grid through time (from Müller et al. 2008). Note that the *grey* polygons correspond to a standard Pacific model of Caribbean evolution (from Ross and Scotese 1988) embedded in the ocean age grid reconstructions

1: **A tectonic setting of two mature, opposed subduction zones**
2: Conjugate shear zone boundaries for the "central" upper plate

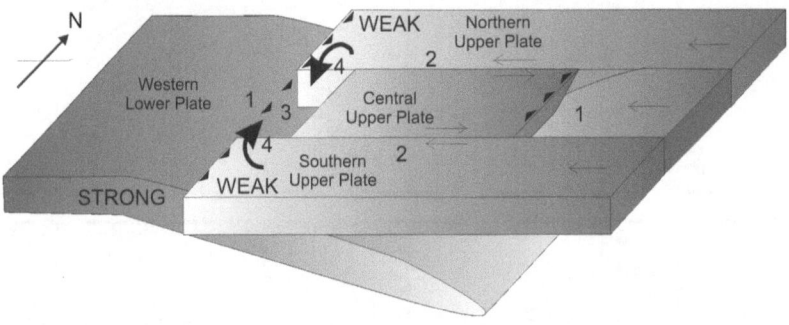

3: Departure of material from the western part of the central upper plate
4: **Earth-radial plate corners** if western lower plate is strong relative to eastern upper plates

Fig. 2.19 Geodynamic hypothesis for the cause of Pirate model kinematics in western Caribbean tectonic evolution. The Caribbean Plate is in a unique geodynamic setting where two subduction zones with opposite polarity and different lengths create unstable triple junctions at the western Caribbean Plate corners. Material can be expected to rotate into the trailing edge of the Caribbean Plate from the north and south, if material to the west and under the western Caribean region is relatively strong and resists deformation

(Keppie 2012; Keppie and Keppie 2012) (Figs. 2.18c–f). Detailed evidence testing this hypothesis is considered elsewhere (Keppie and Keppie 2012). Broadly, the evidence supporting the southeastward escape of the Chortis and Chiapas Blocks from the western Gulf of Mexico include the broadly synchronous, late Cretaceous and the mid-Cenozoic formation of: (1) the Sierra Madre Orientals (Longoria and Suter 1990), (2) the >2 km of excess depth in the western Gulf of Mexico (Feng et al. 1994), and (3) paleocanyons in the southwestern Gulf of Mexico basins (Cantu-Chapa 1987). In the multi-mode model, the missing block of southern Mexico forearc is inferred to have been thrust beneath Mexico by subduction erosion (Keppie et al. 2009, 2012; Fig. 2.16). Back-migrating the Chortis Block in a clockwise manner northwest across the northwest Caribbean Plate corner aligns the sinistral Guayape shear zone (Burkart and Self 1985) with both the northern Caribbean Plate boundary and the Acapulco trench (Fig. 2.12c), suggesting an ephemeral connection. Counter-clockwise rotation of the Chortis Block has lead to re-activation of the Guayape fault dextrally since then (Gordon and Muehlberger 1994).

Diachroneity in the Eocene-Oligocene arc of southern Mexico, older in the west and younger in the east, is traditionally linked to the eastward escape of the Chortis Block from southern Mexico (e.g. Guerrero Garcia and Herrero-Bervera 2010). In contrast, in the multi-mode model, diachroneity may be due to the propagation of the eastern tip of a zone of sinistral transtension eastwards along the northern margin of the missing Mexican forearc (Fig. 2.12c). This hypothesis is compatible with recent data suggesting Eocene-Oligocene arc diachroneity (if any) proceeds rapidly from west-to-east and mostly takes place in the Oligocene (Morán-Zenteno et al. 2009).

2.4.5   Neo-Caribbean Tectonics

At ca. 10 Ma, Caribbean Plate tectonics appear to have changed somewhat from earlier times (Pindell and Barrett 1990; Pindell and Kennan 2009). This last 10 Ma period is called the Neo-Caribbean stage (Pindell and Barrett 1990). Key features of this transition include: emergence of extension in the western Chortis rifts province (Rogers and Mann 2007), emergence of microplate tectonics at the southwest Caribbean Plate corner (Audemard 2009), and a change in the axis of relative rotation between the Caribbean and American Plates (Pindell and Kennan 2009).

In the context of the multi-mode model, some of these changes may have been triggered by a temporary blockage of the microplate channel at the northwest Caribbean Plate corner (Fig. 2.17). There is evidence to suggest that at ca. 11–9 Ma a collision may have occurred between the South Mexico-Chiapas-South Yucatan Blocks and the North Yucatan Block (Kim et al. 2011). The formation of the ca. 11–9 Ma Chiapas fold-and-thrust belt, at the proposed suture, is associated with up to ca. 105 km of shortening (Mandujano-Velazquez and Keppie 2009). A south-dipping slab imaged beneath the Isthmus of Tehuantepec may indicate the closure of a ca. 250 km wide ocean basin (Kim et al. 2011). Microplate collision may be due to the southeastward escape of the South Mexico/Chiapas/South Yucatan Blocks (Andreani et al. 2008a, b), or due to southwestward migration of the North Yucatan Block away from southern Florida (Kim et al. 2011). Such a collision may have temporarily halted y-to-x-directed mass conservation at the northwest Caribbean Plate corner and required the system to conserve mass in other ways.

2.4.6   Meeting of the Americas

The multi-mode model predicts a zone of dextral shear extending southeastwards from the northwest Caribbean Plate corner along the Central American arc (Fig. 1.5f). This can explain dextral shear that exceeds the dextral component of Cocos-Caribbean relative plate motion in the Central American arc (Fig. 2.13d). Similarly, the multi-mode model predicts a zone of sinistral shear extending northwest from the southwest Caribbean Plate corner along the central American arc (Fig. 1.5f), which is consistent with the sinistral displacement in the Romeral fault system at the eastern Choco Block boundary (Ego et al. 1995). Where Pirate tectonics at the northwest and southwest Caribbean Plate corners meet in the central Central American arc, a number of interesting, and previously enigmatic, central Caribbean features have possible explanations. The Hess escarpment appears to occur at the boundary between zones of counter-clockwise and clockwise rotations at the northwest and southwest Caribbean Plate corners, respectively (Fig. 2.1).

Across this boundary, the Managua bookshelf zone (La Femina et al. 2009) may reconcile dextral transtension related to Pirate tectonics at the northwest Caribbean Plate corner with convergence on the North Panama segment of Leeward Antilles subduction (Camacho et al. 2010) related to Pirate tectonics at the southwest Caribbean Plate corner. The join between the North Panama and main Leeward Antilles trench segments appears to occur on the sinistral Romeral Fault (Figs. 2.1, 2.2, 2.3, 2.4, 2.5, 2.6, 2.7). Finally, near the eastern end of the Hess escarpment, the Beata ridge has possibly acommodated and partly formed due to shortening of ambiguous origin (Driscoll and Diebold 1998; DeMets et al. 2007). In the multi-mode model, deformation may be due to the counter-clockwise migration of material into the western Caribbean at the northwest Caribbean Plate corner.

More broadly, counter-clockwise southward escape of North American microplates across the northwest Caribbean Plate corner may be required to explain the inferred, > 500 km Caribbean slab extending beneath northern South America from the Leeward Antilles trench (Fig. 2.10c). Also, the present concave outward curvature of the Cuban reentrant (Fig. 2.1) is currently explained in the Pacific-origin model by counter-clockwise rollback of the Lesser Antilles or Great Arc in the northeast Caribbean during the Cretaceous (Pindell et al. 2005). However, counter-clockwise southward escape of North American microplates across the northwest Caribbean Plate corner would appear to provide an alternate explanation of the Cuban reentrant geometry as a natural corollary of Pirate model tectonics.

2.4.7 Age of Gulf of Mexico Opening

A main prediction of the multi-mode model is that β_N extension in the Gulf of Mexico mostly balances C_E shortening at the eastern Caribbean Plate boundary (Fig. 2.8), and this is consistent with the fact that $\beta_N * h_N$ in the Gulf of Mexico is approximately equal to $C_E * h_E$ in the eastern Caribbean and each exceeds 500,000 km^2 (Fig. 2.1). This main prediction implies that some counter-clockwise opening of the western Gulf of Mexico took place from the late Cretaceous and early Cenozoic (Keppie 2012). This prediction will be tested in subsequent studies.

2.4.8 A Geodynamic Model for the Western Caribbean Region

Changes in the rotation or motion of plates around corner axes tangent to Earth are implied at subduction zones and rifts. Such Earth-tangent plate corners are perhaps a logical consequence of advective cooling of the Earth through mantle convection (Turcotte and Schubert 2002). In the multi-mode model of western Caribbean tectonics, changes in the rotation or motion of plates around axes radial to Earth are inferred at the western Caribbean Plate corners (Fig. 2.19). The existence of Earth-radial plate corners is less clearly a logical consequence of advective cooling of the Earth through mantle convection.

However, secondary processes related to convergent margins formed in the broader convection system may explain changes in the rotation of plates around axes radial to Earth. For example, rotation of forearc microplates are well-observed where indenting elements deform the upper plate at subduction zones (Wallace et al. 2010). However, microplate rotations at the western Caribbean Plate corners are generally not well-explained by indenting elements (Wallace et al. 2010).

A second setting where microplate rotations about axes radial to Earth may occur is where mature, oppositely dipping subduction zones of different axial extent compete at opposite plate boundaries. This is exactly the case in the western Caribbean region (Figs. 1.1, 1.2 and 1.3). This configuration creates sinistral and dextral shear zones at the northern and southern Caribbean Plate boundaries, which propagate west from the Lesser Antilles trench until they approach or meet the Farallon-related trenches in the west (Fig. 2.19). How the northern and southern Caribbean Plate boundaries behave near the western Caribbean Plate corners is dependent on the relative strength of material to the west, under, and to the north and south of the western Caribbean region. If material to the west is relatively weak, it might tear and a Pacific model of tectonics may result. If the material at depth is relatively weak, it might tear and/or melt and an in-situ model of tectonics may result. If the material to the north and south is relatively weak, it might tear and rotate into the trailing edge of the Caribbean Plate as in the Pirate tectonic model.

Farallon-related lithosphere lies to the west and under the western Caribbean region and is likely the most significant regional rheological contrast. Interpretation of Pirate model tectonics and Earth radial plate corners at the western Caribbean plate corners, in general, thus indicates the relative strength of Farallon-related lithosphere in the western Caribbean context (Fig. 2.19).

2.5 Conclusions

The tectonic evolution of the Gulf of Mexico and the western Caribbean Plate region has been evaluated with respect to the theoretical principles of global surface mass conservation, local mass conservation, and triple point stability. The three end-member models possible for western Caribbean tectonics all involve material moving out of the western Caribbean region eastwards relative to North and South America: (1) the Pacific model involves material moving into the western Caribbean region from the west; (2) the In-situ model involves material moving into the western Caribbean region from depth; (3) the Pirate model involves material moving into the western Caribbean region from the north and south.

Analysis of the geological record of the Gulf of Mexico and the Caribbean regions indicates that the Pacific model may have operated during the Cretaceous; the In-situ model probably operated in the Cretaceous and Late Miocene-Present; the Pirate model has been dominant in most of the Cenozoic. Since the late Cretaceous, North American-derived microplates may have rotated counter-clockwise into the western Caribbean Plate region from the north. Since the early Eocene, the South

American Plate has rotated clockwise into the western Caribbean Plate region from the south. This multi-model genesis for the Caribbean Plate and the Gulf of Mexico presents a testable paleogeographic model to be developed (Fig. 2.17) that satisfies most existing data. Key testable predictions include: (1) a Cretaceous position of the Chortis Block within the western Gulf of Mexico (Fig. 2.18), (2) a component of dextral N–S motion that contributed to the formation of, and bend in, the Sierra Madre Orientals orogeny in eastern Mexico, and (3) a component of Gulf of Mexico opening that is late Cretaceous to mid-Cenozoic in age. Greater rheological strength of Farallon-related lithosphere relative to North, Central, and South American lithosphere may have caused the Pirate model to occur (Fig. 2.19).

References

Alzaga-Ruiz H, Granjeon D, Lopez M, Séranne M, Roure F (2009a) Gravitational collapse and Neogene sediment transfer across the western margin of the Gulf of Mexico: insights from numerical models. Tectonophysics 470(1–2):21–41

Alzaga-Ruiz H, Lopez M, Roure F, Séranne M (2009b) Interactions between the Laramide Foreland and the passive margin of the Gulf of Mexico: tectonics and sedimentation in the Golden Lane area, Veracruz State, Mexico. Mar Petrol Geol 26(6):951–973

Andreani L, Le Pichon X, Rangin C, Martínez-Reyes J (2008a) The southern Mexico block: main boundaries and new estimation for its Quaternary motion. Bulletin de la Societe Geologique de France 179(2):209–223

Andreani L, Rangin C, Martínez-Reyes J, Le Roy C, Aranda-García M, Le Pichon X, Peterson-Rodriguez R (2008b) The Neogene Veracruz fault: evidences for left-lateral slip along the southern Mexico block. Bulletin de la Societe Geologique de France 179(2):195–208

Audemard FE, Audemard FA (2002) Structure of the Mérida Andes, Venezuela: relations with the South America-Caribbean geodynamic interaction. Tectonophysics 345(1–4):299–327

Audemard FA (2009) Key issues on the post-Mesozoic southern Caribbean Plate boundary. Geol Soc Lond Spec Pub 328(1):569–586

Audemard FA, Romero G, Rendon H, Cano V (2005) Quaternary fault kinematics and stress tensors along the southern Caribbean from fault-slip data and focal mechanism solutions. Earth-Sci Rev 69(3–4):181–233

Authemayou C, Brocard G, Teyssier C, Simon-Labric T, Guttiérrez A, Chiquín EN, Morán S (2011) The Caribbean-North America-Cocos triple junction and the dynamics of the Polochic-Motagua fault systems: pull-up and zipper models. Tectonics 30 (TC 3010). doi:10.1029/2010TC002814

Backé G, Dhont D, Hervouët Y (2006) Spatial and temporal relationships between compression, strike-slip and extension in the central Venezuelan Andes: clues for Plio-Quaternary tectonic escape. Tectonophysics 425(1–4):25–53

Ball MM, Harris CGA, Supko PR (1969) Atlantic opening and the origin of the Caribbean. Nature 223(5202):167–168

Bandy WL, Hilde TWC, Yan CY (2000) The Rivera-Cocos plate boundary: implications for Rivera-Cocos relative motion and plate fragmentation. Geol Soc Am Spec Pap 334:1–28

Bird DE, Burke K, Hall SA, Casey JF (2005) Gulf of Mexico tectonic history: hotspot tracks, crustal boundaries, and early salt distribution. Am Assoc Pet Geol Bull 89(3):311–328

Bouysse P (1988) Opening of the Grenada back-arc basin and evolution of the Caribbean Plate during the Mesozoic and early Paleogene. Tectonophysics 149(1–2):121–143

Bouysse P, Westercamp D (1990) Subduction of Atlantic aseismic ridges and Late Cenozoic evolution of the Lesser Antilles island arc. Tectonophysics 175(4):349–380

Burkart B (1978) Offset across the Polochic fault of Guatemala and Chiapas, Mexico. Geology 6(6):328–332

Burkart B, Self S (1985) Extension and rotation of crustal blocks in northern Central America and effect on the volcanic arc. Geology 13(1):22–26

Burke K (1988) Tectonic evolution of the Caribbean. Annu Rev Earth Planet Sci 16(1):201–230

Camacho E, Hutton W, Pacheco JF (2010) A new look at evidence for a Wadati-Benioff zone and active convergence at the North Panama deformed belt. Bull Seismol Soc Am 100(1):343–348

Campos-Enriquez J, Alatorre-Zamora M (1998) Shallow crustal structure of the junction of the grabens of Chapala, Tepic-Zacoalco and Colima, Mexico. Geofisica Internacional 37(4):263–282

Cantu-Chapa A (1987) The Bejuco paleocanyon (Cretaceous-Paleocene) in the Tampico district, Mexico. J Pet Geol 10(2):207–218

Chapa AC (1985) Is there a Chicontepec paleocanyon in the Paleogene of eastern Mexico? J Pet Geol 8(4):423–434

Clark K, Foster C, Damon P (1982) Cenozoic mineral deposits and subduction-related magmatic arcs in Mexico. Geol Soc Am Bull 93(6):533–544

Cox A, Hart RB (1986) Plate tectonics: how it works. Wiley-Blackwell, Palo Alto, p 391

DeMets C, Mattioli G, Jansma P, Rogers R, Tenorio C, Turner HL (2007) Present motion and deformation of the Caribbean plate: constraints from new GPS geodetic measurements from Honduras and Nicaragua. Geol Soc Am Spec Pap 428:21–36

DeMets C, Gordon RG, Argus DF (2010) Geologically current plate motions. Geophys J Int 181(1):1–80

Dewey JF, Burke K (1980) Episodicity, sequence, style at convergent plate boundaries. In the continental crust and its mineral deposits. Geol Assoc Can Spec Pap 20:553–573

Dickinson WR (2009) The Gulf of Mexico and the southern margin of Laurentia. Geology 37(5):479–480

Dickinson WR, Snyder WS (1979a) Geometry of subducted slabs related to San Andreas transform. J Geol 87(6):609–627

Dickinson WR, Snyder WS (1979b) Geometry of triple junctions related to San Andreas transform. J Geophys Res 84(B2):561–572

Dixon TH, Farina F, Demets C, Jansma P, Mann P, Calais E (1998) Relative motion between the Caribbean and North American Plates and related boundary zone deformation from a decade of GPS observations. J Geophys Res 103(B7):15157–15182

Driscoll NW, Diebold JB (1998) Deformation of the Caribbean region: one plate or two? Geology 26(11):1043–1046

Ego F, Ansan V (2002) Why is the Central Trans-Mexican Volcanic Belt (102°–99°W) in transtensive deformation? Tectonophysics 359(1–2):189–208

Ego F, Sébrier M, Yepes H (1995) Is the Cauca-Patia and Romeral fault system left or right lateral? Geophys Res Lett 22(1):33–36

English J, Johnston S (2004) The Laramide orogeny: what were the driving forces? Int Geol Rev 46(9):833

English JM, Johnston ST, Wang K (2003) Thermal modelling of the Laramide orogeny: testing the flat-slab subduction hypothesis. Earth Planet Sci Lett 214(3–4):619–632

Escalona A, Mann P (2011) Tectonics, basin subsidence mechanisms, and paleogeography of the Caribbean-South American Plate boundary zone. Mar Petrol Geol 28:8–39

ESRI (2011) A zipped shapefile of tectonic plate boundaries for the world. Includes boundary type attribute. It is intended for educational use. Last modified October 5, 2011. Downloaded from http://www.arcgis.com/home/item.html?id=357b0e32423f43cebf9f844ae70f7d1c

Farr TG et al (2007) The shuttle radar topography mission. Rev Geophys 45:33. doi:10.1029/2005RG000183

Feng J, Buffler RT, Kominz MA (1994) Laramide orogenic influence on late Mesozoic-Cenozoic subsidence history, western deep Gulf of Mexico basin. Geology 22(4):359–362

Fillon RH (2007) Mesozoic Gulf of Mexico basin evolution from a planetary perspective and petroleum system implications. Pet Geosci 13(2):105–126

Flotte N, Martinez-Reyes J, Rangin C, Le Pichon X, Husson L, Tardy M et al (2008) The Rio
    Bravo fault, a major late Oligocene left-lateral shear zone. Bulletin de la Societe Geologique
    de France 179(2):147–160
Giunta G, Beccaluva L (2006) Caribbean plate margin evolution: constraints and current problems.
    Geoligica Acta 4(1–2):265–277
Gordon MB, Muehlberger WR (1994) Rotation of the Chortís block causes dextral slip on the
    Guayape fault. Tectonics 13(4):858–872
Gorney D, Escalona A, Mann P, Magnani MB, Bolivar Study Group (2007) Chronology of Ce-
    nozoic tectonic events in western Venezuela and the Leeward Antilles based on integration of
    offshore seismic reflection data and on-land geology. Am Assoc Pet Geol Bull 91(5):653–684
Gose WA (1985) Paleomagnetic results from Honduras and their bearing on Caribbean tectonics.
    Tectonics 4(6):565–585
Gough DI, Heirtzler JR (1969) Magnetic anomalies and tectonics of the Cayman Trough. Geophys
    J Int 18(1):33–49
Griffin WR, Foland KA, Stern RJ, Leybourne MI (2010) Geochronology of bimodal alkaline vol-
    canism in the Balcones Igneous Province, Texas: implications for Cretaceous intraplate mag-
    matism in the Northern Gulf of Mexico Magmatic Zone. J Geol 118(1):1–21
Guerrero-Garcia JC, Herrero-Bervera E (2010) Tectonics of southwestern Mexico, isotopic evi-
    dence, nuclear Central America, Late Cretaceous break up. Studia Geophysica et Geodaetica
    54(3):403–415
Gurnis M, Hager BH (1988) Controls of the structure of subducted slabs. Nature 335(6188):317–
    321
Gutscher M, Malavieille J, Lallemand S, Collot J-Y (1999) Tectonic segmentation of the North An-
    dean margin: impact of the Carnegie Ridge collision. Earth Planet Sci Lett 168(3–4):255–270
Guzmán-Speziale M (2009) A seismotectonic model for the Chortis Block. Geol Soc Lond Spec
    Pub 328(1):197–204
Guzmán-Speziale M (2010) Beyond the Motagua and Polochic faults: active strike-slip faulting
    along the western North America-Caribbean Plate boundary zone. Tectonophysics 496(1–
    4):17–27. doi:10.1016/j.tecto.2010.10.002
Guzmán-Speziale M, Meneses-Rocha JJ (2000) The North America-Caribbean Plate boundary
    west of the Motagua-Polochic fault system: a fault jog in southeastern Mexico. J S Am Earth
    Sci 13(4–5):459–468
Guzmán-Speziale M, Pennington WD, Matumoto T (1989) The triple junction of the North Amer-
    ica, Cocos, and Caribbean Plates: seismicity and tectonics. Tectonics 8(5):981–997
Hall ML, Wood CA (1985) Volcano-tectonic segmentation of the northern Andes. Geology
    13(3):203–207
Hippolyte J, Mann P (2010) Neogene-Quaternary tectonic evolution of the Leeward Antilles is-
    lands (Aruba, Bonaire, Curaçao) from fault kinematic analysis. Mar Petrol Geol 28:259–277
Holcombe T, Vogt PR, Matthews JE, Murchison RR (1973) Evidence for sea-floor spreading in the
    Cayman Trough. Earth Planet Sci Lett 20(3):357–371
Hossack J (1994) The age of salt in the Gulf of Mexico basin - comment. J Pet Geol 17(3):351–354
Husson L, Henry P, Le Pichon X (2008a) Thermal regime of the NW shelf of the Gulf of Mexico.
    Part A: thermal and pressure fields. Bulletin de la Societe Geologique de France 179(2):129–137
Husson L, Le Pichon X, Henry P, Flotte N, Rangin C (2008b) Thermal regime of the NW
    shelf of the Gulf of Mexico. Part B: heat flow. Bulletin de la Societe Geologique de France
    179(2):139–145
James KH (2006) Arguments for and against the Pacific origin of the Caribbean Plate: discussion,
    finding for an inter-American origin. Geologica Acta 4(1–2):279–302
James KH (2009) Evolution of Middle America and the in situ Caribbean Plate model. Geol Soc
    Lond Spec Pub 328(1):127–138
James KH (2009) In situ origin of the Caribbean: discussion of data. Geol Soc Lond Spec Pub
    328(1):77–125

James KH (2013) Caribbean geology: extended and subsided continental crust sharing history eastern North America, the Gulf of Mexico, the Yucatan Basin and northern South America. Geosci Canada 40(1). doi:http://dx.doi.org/10.12789%2Fgeocanj%2F2013.40.001

Jones RR, Holdsworth RE, Bailey W (1997) Lateral extrusion in transpression zones: the importance of boundary conditions. J Struct Geol 19(9):1201–1217

Kennan L, Pindell JL (2009) Dextral shear, terrane accretion and basin formation in the Northern Andes: best explained by interaction with a Pacific-derived Caribbean Plate? Geol Soc Lond Spec Pub 328(1):487–531

Keppie DF (2012) Derivation of the Chortis and Chiapas blocks from the western Gulf of Mexico in the latest Cretaceous-Cenozoic: the Pirate model. Int Geol Rev 54(15):1765–1775. doi:10.1 080/00206814.2012.676356

Keppie DF (2013) The rationale and essential elements for the new Pirate model of Caribbean tectonics. Geosci Can 40(1). doi:http://dx.doi.org/10.12789%2Fgeocanj%2F2013.40.002

Keppie DF, Keppie JD (2012) An alternative Pangea reconstruction for MiddleAmerica with the Chortis Block in the Gulf of Mexico: tectonic implications. Int Geol Rev 54(14):1685–1696. doi:10.1080/00206814.2012.676361

Keppie JD, Morán-Zenteno D (2005) Tectonic implications of alternative Cenozoic reconstructions for southern Mexico and the Chortis Block. Int Geol Rev 47(5):473–491

Keppie DF, Currie CA, Warren CJ (2009) Subduction erosion models: comparing finite element numerical models with the geological record. Earth Planet Sci Lett 287(1–2):241–254

Keppie JD, Morán-Zenteno DJ, Martiny B, González-Torres E (2009) Synchronous 29–19 Ma arc hiatus, exhumation and subduction of forearc in southwestern Mexico. Geol Soc Lond Spec Pub 328(1):169–179

Keppie DF, Hynes AJ, Lee JKW, Norman M (2012) Oligocene-Miocene back-thrusting in southern Mexico linked to the rapid subduction erosion of a large forearc block. Tectonics 31 (TC2008). doi:10.1029/2011TC002976

Kerr AC, Tarney J, Marriner GF, Nivia A, Saunders AD (1997) The Caribbean-Colombian Cretaceous Igneous Province: the internal anatomy of an oceanic plateau. In: Mahoney JJ, Coffin MF (eds) Large igneous provinces: Continental, oceanic, and planetary flood volcanism, vol 100. American Geophysical Union geophysical monograph, Washington, DC, pp 123–144

Kerr AC, Tarney J (2005) Tectonic evolution of the Caribbean and northwestern South America: the case for accretion of two Late Cretaceous oceanic plateaus. Geology 33(4):269–272

Kim Y, Clayton R, Keppie DF (2011) Evidence of a collision between the Yucatan Block and Mexico in the Miocene. Geophys J Int. doi:10.1111/j.1365-246X.2011.05191.x

Kuhn TS (ed) (1977) Objectivity, value judgement, and theory choice. In: The essential tension. University of Chicago Press, Chicago, pp 320–339

La Femina PC, Dixon TH, Strauch W (2002) Bookshelf faulting in Nicaragua. Geology 30(8):751–754

La Femina PC, Dixon TH, Govers R, Norabuena E, Turner H, Saballos A, Mattioli G, Protti M, Strauch W (2009) Fore-arc motion and Cocos Ridge collision in Central America. Geochem Geophys Geosys 10(5). doi:10.1029/2008GC002181

Labails C, Olivet JL, Aslanian D, Roest WR (2010) An alternative early opening scenario for the Central Atlantic Ocean. Earth Planet Sci Lett 297(3–4):355–368. doi:10.1016/j. epsl.2010.06.024

Lang H, Frerichs W (1998) New Planktic Foraminiferal Data Documenting Coniacian Age for Laramide Orogeny Onset and paleooceanography in Southern Mexico. J Geol 106(5):635–640

Leroy S, Mauffret A, Patriat P, Mercier deLB (2000) An alternative interpretation of the Cayman trough evolution from a reidentification of magnetic anomalies. Geophys J Int 141:539–557

Li C, van der Hilst RD, Engdahl ER, Burdick S (2008) A new global model for P wave speed variations in Earth's mantle. Geochem Geophys Geosys 9:21. doi:10.1029/2007GC001806

Longoria J, Suter M (1990) Structural traverse across the Sierra Madre oriental fold-thrust belt in east-central Mexico: alternative interpretation and reply. Geol Soc Am Bull 102(2):261–266

Lyon-Caen H et al (2006) Kinematics of the North American-Caribbean-Cocos plates in Central America from new GPS measurements across the Polochic-Motagua fault system. Geophys Res Lett 33:5

Macdonald R, Hawkesworth CJ, Heath E (2000) The Lesser Antilles volcanic chain: a study in arc magmatism. Earth-Sci Rev 49(1–4):1–76

MacRae G, Watkins JS (1992) Evolution of the Destin Dome, offshore Florida, north-eastern Gulf of Mexico. Mar Petrol Geol 9(5):501–509

Mandujano-Velazquez JJ, Keppie JD (2009) Middle Miocene Chiapas fold and thrust belt of Mexico: a result of collision of the Tehuantepec Transform/Ridge with the Middle America Trench. Geol Soc Lond Spec Pub 327(1):55–69

Mann P (2007) Overview of the tectonic history of northern Central America. Geol Soc Am Spec Pap 428:1–19

Marquez-Azua B, DeMets C (2009) Deformation of Mexico from continuous GPS from 1993 to 2008. Geochem Geophys Geosys 10:16

Marsaglia KM, Davis AS, Rimkus K, Clague DA (2006) Evidence for interaction of a spreading ridge with the outer California borderland. Mar Geol 229(3–4):259–272

Marton G, Buffler RT (1994) Jurassic reconstruction of the Gulf of Mexico Basin. Int Geol Rev 36:545–586

Maus SU, Barckhausen H, Berkenbosch N, Bournas J, Brozena V, Childers F, Dostaler JD, Fairhead C, Finn RRB, von Frese C, Gaina S, Golynsky R, Kucks H, Luhr P, Milligan S, Mogren D, Muller O, Olesen M, Pilkington R, Saltus B, Schreckenberger E, Thebault F, Caratori T (2010) EMAG2: A 2-arc-minute resolution Earth Magnetic Anomaly Grid compiled from satellite, airborne and marine magnetic measurements. Geochem Geophys Geosyst 10(8). doi:10.1029/2009GC002471

McKenzie DP, Morgan WJ (1969) Evolution of triple junctions. Nature 224(5215):125–133

Meschede M, Frisch W (1998) A plate-tectonic model for the Mesozoic and Early Cenozoic history of the Caribbean plate. Tectonophysics 296(3–4):269–291

Michaud F, Witt C, Royer J (2009) Influence of the subduction of the Carnegie volcanic ridge on Ecuadorian geology: reality and fiction. Geol Soc Am Memoir 204:217–228

Mickus K, Stern RJ, Keller GR, Anthony EY (2009) Potential field evidence for a volcanic rifted margin along the Texas Gulf Coast. Geology 37(5):387–390

Molnar P, Sykes LR (1969) Tectonics of the Caribbean and Middle America Regions from focal mechanisms and seismicity. Geol Soc Am Bull 80(9):1639–1684

Montes C, Hatcher J, Restrepo-Pace PA (2005) Tectonic reconstruction of the northern Andean blocks: oblique convergence and rotations derived from the kinematics of the Piedras-Girardot area, Colombia. Tectonophysics 399(1–4):221–250

Montes C, Guzman G, Bayona G, Cardona A, Valencia V, Jaramillo C (2010) Clockwise rotation of the Santa Marta massif and simultaneous Paleogene to Neogene deformation of the Plato-San Jorge and Cesar-Ranchería basins. J S Am Earth Sci 29:832–848. doi:10.1016/j.jsames.2009.07.010

Morán-Zenteno DJ, Keppie JD, Martiny B, González-Torres E (2009) Reassessment of the Paleogene position of the Chortis Block relative to southern Mexico: hierarchichal ranking of data and features. Revista Mexicana de Ciencias Geologicas 26(1):177–188

Morgan WJ (1968) Rises, trenches, great faults, and crustal blocks. J Geophys Res 73(6):1959–1982. doi: 10.1029/JB073i006p01959

Morgan JP, Ranero C, Vannucchi P (2008) Intra-arc extension in Central America: links between plate motions, tectonics, volcanism, and geochemistry. Earth Planet Sci Lett 272(1–2):365–371

Müller RD, Sdrolias M, Gaina C, Roest WR (2008) Age, spreading rates, and spreading asymmetry of the world's ocean crust. Geochem Geophys Geosys 9 (Q04006). doi:10.1029/2007GC001743

Nagihara S, Jones KO (2005) Geothermal heat flow in the northeast margin of the Gulf of Mexico. Am Assoc Pet Geol Bull 89(6):821–831

Oskin M, Stock J, Martin-Barajas A (2001) Rapid localization of Pacific-North America plate motion in the Gulf of California. Geology 29(5):459–462

Pindell JL (1985) Alleghenian reconstruction and subsequent evolution of the Gulf of Mexico, Bahamas and proto-Caribbean. Tectonics 4(1):1–39

Pindell JL, Cande SC, Pitman WCIII, Rowley DB, Dewey JF, LaBrecque J, Haxby W (1988) A plate-kinematic framework for models of Caribbean evolution. Tectonophysics 155(1–4):121–138

Pindell JL (2010) Alleghenian reconstruction and subsequent evolution of the Gulf of Mexico, Bahamas, and proto-Caribbean. Tectonics 4(1):1–39. doi:10.1029/TC004i001p00001

Pindell JL, Barrett S (1990) Geological evolution of the Caribbean region; a plate tectonic perspective. In: Dengo GA, Case JE (eds) The geology of North America. Geological Society of America, Boulder, pp 405–432

Pindell JL, Dewey JF (1982) Permo-Triassic reconstruction of western Pangea and the evolution of the Gulf of Mexico/Caribbean region. Tectonics 1(2):179–211

Pindell JL, Kennan L (2009) Tectonic evolution of the Gulf of Mexico, Caribbean and northern South America in the mantle reference frame: an update. Geol Soc Lond Spec Pub 328(1):1–55

Pindell JL, Kennan L, Maresch WV, Stanek K-P, Draper G, Higgs R (2005) Plate-kinematics and crustal dynamics of circum-Caribbean arc-continent interactions: tectonic controls on basin development in Proto-Caribbean margins. Geol Soc Am Spec Pap 394:7–52

Pindell JL, Kennan L, Stanek K-P, Maresch WV, Draper G (2006) Foundations of Gulf of Mexico and Caribbean evolution: eight current controversies resolved. Geologica Acta 4(1–2):303–341

Rangin C, Le Pichon X, Flotte N, Husson L (2008a) Cenozoic gravity tectonics in the northern Gulf of Mexico induced by crustal extension. A new interpretation of multichannel seismic data. Bulletin de la Societe Geologique de France 179(2):117–128

Rangin C, Le Pichon X, Martinez-Reyes J, Aranda-Garcia M (2008b) Gravity tectonics and plate motions: the western margin of the Gulf of Mexico Introduction. Bulletin de la Societe Geologique de France 179(2):107–116

Ratschbacher L, Franz L, Min M, Bachmenn R, Martens U, Stanek K, Stubner K, Nelson BK, Herrmann U, Weber B, López-Martínez M, Jonckheere R, Sperner B, Tichomirowa M, Mc-Williams MO, Gordon M, Meschede M, Bock P (2009) The North American-Caribbean Plate boundary in Mexico-Guatemala-Honduras. Geol Soc Lond Spec Pub 328(1):219–293

Rea DK, Malfait BT (1974) Geologic evolution of the northern Nazca Plate. Geology 2(7):317–320

Reed JM (1994) Probable Cretaceous-to-recent rifting in the Gulf of Mexico basin: an answer to Callovian salt deformation and distribution problems? Part 1. J Pet Geol 17(4):429–444

Reed JM (1995) Probable Cretaceous-to-recent rifting in the Gulf of Mexico basin: an answer to Callovian salt deformation and distribution problems? Part 2. J Pet Geol 18(1):49–74

Rodriguez M, DeMets C, Rogers R, Tenorio C, Hernandez D (2009) A GPS and modelling study of deformation in northern Central America. Geophys J Int 178(3):1733–1754

Rogers RD, Mann P (2007) Transtensional deformation of the western Caribbean-North America plate boundary zone. Geol Soc Am Spec Pap 428:37–64

Rogers RD, Mann P, Emmet PA (2007) Tectonic terranes of the Chortis Block based on integration of regional aeromagnetic and geologic data. Geol Soc Am Spec Pap 428:65–88

Rosencrantz E, Ross MI, Sclater JG (1988) Age and spreading history of the Cayman Trough as determined from depth, heat flow, and magnetic anomalies. J Geophys Res 93(B3):2141–2157

Rosencrantz E, Sclater JG (1986) Depth and age in the Cayman Trough. Earth Planet Sci Lett 79(1–2):133–144

Ross MI, Scotese CR (1988) A hierarchical tectonic model of the Gulf of Mexico and Caribbean region. Tectonophysics 155(1–4):139–168

Salvador A (1987) Late Triassic-Jurassic paleogeography and origin of Gulf of Mexico Basin. AAPG Bulletin 71(4):419–451

Sandwell DT, Smith WHF (2009) Global marine gravity from retracked Geosat and ERS-1 altimetry ridge segmentation versus spreading rate. J Geophys Res 114 (B01411). doi:10.1029/2008JB006008

Sawyer D, Buffler RT, Pilger R (1991) The crust under the Gulf of Mexico Basin. In: Salvador A (ed) The geology of North America. Geological Society of America, Boulder, pp 53–72

Seton M, Muller RD, Zahirovic S, Gaina C, Torsvik TH, Shephard G, Talsma A, Gurnis M, Turner M, Maus S, Chandler M (2012) Global continental and ocean basin reconstructions since 200 Ma. Earth Sci Rev 112(3–4):212–270. doi:10.1016/j.earscirev.2012.03.002

Silva-Romo G (2008) Guayape-Papalutla fault system: a continuous Cretaceous structure from southern Mexico to the Chortís block? Tectonic implications. Geology 36(1):75–78

Silver EA, Case JE, MacGillavry HJ (1975) Geophysical study of the Venezuelan Borderland. Geol Soc Am Bull 86(2):213–226

Spencer A (1969) Alkalic igneous rocks of the Balcones Province, Texas. Petrology 10(2):272–306

Stein CA, Stein S (1992) A model for the global variation in oceanic depth and heat flow with lithospheric age. Nature 359(6391):123–129

Suter F, Sartori M, Neuwerth R, Gorin G (2008) Structural imprints at the front of the Chocó-Panamá indenter: field data from the North Cauca Valley Basin, Central Colombia. Tectonophysics 460(1–4):134–157

Talavera-Mendoza O, Ruiz J, Corona-Chavez P, Gehrels GE, Sarmiento-Villagrana A, Garcia-Diaz JL, Salgado-Souto SA (2013) Origin and provenance of basement metasedimentary rocks from the Xolapa Complex: new constraints on the Chortis-southern Mexico connection. Earth Planet Sci Lett 369–370:188–199. doi: 10.1016/j.epsl.2013.03.021

Talbot C (2004) Extensional evolution of the Gulf of Mexico basin and the deposition of Tertiary evaporites - discussion. J Pet Geol 27(1):95–104

Torres-de Leon R, Solari L, Ortega-Gutierrez F, Martens U (2012) The Chortis Block - southwestern Mexico connections: U-Pb zircon geochronology constraints. Am J Sci 312(3):288–313. doi: 10.2475/03.2012.02

Trenkamp R, Kellogg JN, Freymueller JT, Mora HP (2002) Wide plate margin deformation, southern Central America and northwestern South America, CASA GPS observations. J S Am Earth Sci 15(2):157–171

Turcotte DL, Schubert G (2002) Geodynamics. Cambridge University Press, Cambridge

Vallejo C, Winkler W, Spiking RA, Luzieux L, Heller F, Bussy F (2009) Model and timing of terrane accretion in the forearc of the Andes in Ecuador. Geol Soc Am Mem 204:197–216

Valls Alvarez RA (2009) Geological evolution of the NW corner of the Caribbean Plate. Geol Soc Lond Spec Pub 328(1):205–217

Wallace LM, Ellis S, Mann P (2010) Collisional model for rapid fore-arc block rotations, arc curvature, and episodic back-arc rifting in subduction settings. Geochem Geophys Geosys 10 (Q05001). doi:10.1029/2008GC002220

Walper J, Rowett CL (1972) Plate tectonics and origin of Caribbean Sea and Gulf of Mexico. GCAGS 22:105–116

Wark DA, Kempter KA, McDowell FW (1990) Evolution of waning, subduction-related magmatism, northern Sierra Madre Occidental, Mexico. Geol Soc Am Bull 102(11):1555–1564

Wilson HH (1993) The age of salt in the Gulf of Mexico basin. J Pet Geol 16(2):125–151

Wilson HH (2003) Extensional evolution of the Gulf of Mexico Basin and the deposition of tertiary evaporites. J Pet Geol 26(4):403–428

Wilson HH (2004) Extensional evolution of the Gulf of Mexico Basin and the depostion of tertiary evaporites. J Pet Geol 27(1):105–110

Yu Z, Lerche I, Lowrie A (1992) Thermal impact of salt: simulation of thermal anomalies in the Gulf of Mexico. Pure Appl Geophys 138(2):181–192

Chapter 3
Normalization Analysis for Possibly-Unstable Triple Junction Zones

Abstract This chapter demonstrates the need for multiple working hypotheses in the evaluation of complex tectonic zones such as the diffuse northwest Caribbean triple junction zone. Where an underlying incompatibility can be identified in the contextual major plate system, a number of nonrigid processes may arise in isolation or in some combination to stabilize the system. Recognition of the relative roles for the various processes requires explicit consideration of all of these possibilities. Characteristics of the different processes are also not always diagnostic. In this case, it is a combination of phenomena preserved in the rock record that must be interpreted together in order to discriminate between the different possibilities.

3.1 Introduction

There are several ways to approach modeling the long-term evolution of the solid Earth. A fundamental geodynamic approach would take complete knowledge of material properties such as density, rheology, and thermal parameters (expansivity, conductivity, etc.) and complete knowledge of governing physical equations (e.g., conservation of mass, momentum, and energy and applicable constitutive equations in a classical mechanical treatment) to predict the kinematic evolution of the solid Earth through time (Cloetingh and Negendank 2010). In practice, however, our knowledge of the material properties is generally insufficient to reproduce the specific evolution of the solid Earth when compared against models derived directly from interpretations of the rock record. Thus, reconstructing earth evolution from direct observation(s) and interpretation(s) of the rock record remains a vital methodology (Seton et al. 2012). Direct kinematic reconstructions provide a means to test geodynamic predictions, but may also be preferable where the accurate kinematic constraints provide critical boundary conditions for other problems such as, climate and ocean change, exploration for natural resources, and management of natural hazards.

Direct kinematic reconstructions of solid Earth evolution are not trivial to achieve; however, even once this approach is prioritized. In an ideal scenario, a dense field of passive point markers could be inserted into the Earth system and monitored through time to achieve a comprehensive constraint. However, such complete and regular data are largely impossible for locations at depth and for times

D. Fraser Keppie, *The Analysis of Diffuse Triple Junction Zones in Plate Tectonics and the Pirate Model of Western Caribbean Tectonics,* SpringerBriefs in Earth Sciences, DOI 10.1007/978-1-4614-9616-8_3, © Springer Science+Business Media New York 2014

in the geological past, given the practical issues confronted. The observation and interpretation of the rock record is fundamentally limited by access and preservation issues, as well as intrinsic ambiguity in the results of various analytical methods. The situation is better for modern tectonics where global positioning system (GPS) stations and seismic monitoring can yield more complete constraints on regional strain, at least at the surface of Earth or within the upper lithosphere. However, this paper is largely concerned with reconstructing the past for which such real-time observations are unavailable.

In this chapter, I present a specific approach to aid the evaluation of complex triple junctions where deformation may be distributed and heterogeneous. Deformation across such zones can be caused by local changes to boundary conditions or as a regional response to an underlying incompatibility in the major plate motions and plate boundary orientations. The normalization process outlined here provides one way to identify where major plate motions and plate boundaries may be incompatible at their triple junction zones and what general nonrigid processes can lead to further evolution of the physical system in question. In general, possible solutions to the identified incompatibilities are non-unique, but the identification of the end-member possibilities is prerequisite for robust evaluations of the rock record.

The normalization process is as follows. Plate boundary zones that are known to host distributed and heterogeneous deformation can be normalized in terms of linear, infinitesimally-wide classic plate boundary types, i.e., the simple plate boundary types that would be the case were deformation obeying classic plate tectonics. These simple plate boundary approximations can be deduced from the relative motions between adjacent major plates and the general orientations of the intervening boundaries. The normalized boundaries can then be analyzed for stability where they intersect in hypothetical triple junctions (McKenzie and Morgan 1969). If the hypothetical triple junctions are unstable, then the distributed and heterogeneous deformation removed during the normalization analysis may reflect the system's attempt to accommodate the underlying instability. In general, there are four ways for the system to respond to a normalized incompatibility, all of which involve nonrigid events or processes that do not conform to classic plate tectonic systems in a steady-state. These are: (1) plate amalgamation events, (2) plate breakup events, (3) plate relative velocity change events, and (4) lateral tectonic events, in which finite-width regions of the surface of Earth accommodate ongoing deformation about vertical axes (e.g., microplate escape/capture processes).

The value of the normalization analysis outlined here is that it becomes possible to investigate features of the rock record in terms of these four end-member styles of non-rigid process. Without such an analysis, the underlying incompatibility between the major plate motions may not be recognized, in which case the systematic nature of the known non-rigidity may not be recognized either. Instead, deviations from classic plate tectonics may be attributed only to local boundary condition effects, such as the collision of indenting elements at a convergent margin. The normalization analysis outlined here can be easily adapted and applied to arbitrary apparent triple junction configurations as needed. However, in the context of the present work and for demonstration purposes here, I illustrate and discuss one such analysis in some detail with regards to a specific triple junction configuration similar to the northwest Caribbean Plate corner.

The normalization analysis follows four basic steps. First, define a movable, doubly-tangent Euclidean reference frame, R, in which it is possible to calculate and compare relative velocities and displacements both at specific points of interest and along either rigid or nonrigid flow lines. Second, construct a reference model, M, in R for the triple junction configuration implied by the relative major plate motions and the trends of their common boundaries. This is done by normalizing deformation between major plates, which may be complex, distributed, and heterogeneous, on infinitesimally-wide boundary lines corresponding to classic plate boundary types. Third, evaluate M in R using the stability analysis pioneered by McKenzie and Morgan (1969). If M is stable, one can assume that all complex, distributed, and heterogeneous deformation is related to local boundary condition parameters only. If M is unstable, complex, distributed, and heterogeneous deformation between adjoining plates may reflect a systematic system response to the instability or incompatibility identified in the normalization. Fourth, construct qualitatively the end-member scenarios in which an unstable M can be made stable via a systematic system response. As noted, these include plate amalgamation events, plate breakup events, plate velocity change events, or lateral tectonic events, either alone or in some combination.

3.2   Step 1: Define a Doubly-Tangent, Moving Euclidean Tectonic Reference Frame

The surface expression of terrestrial tectonics takes place on the surface of Earth. Geometrically, the surface of Earth is a topographically-variable ellipsoid in detail, but is commonly treated simply as a sphere. In either case, the underlying geometry of terrestrial tectonics is non-Euclidean. Furthermore, although the conventional geographic graticule (i.e., the grid of lines of latitude and longitude defined relative to the spin axis of Earth) provides a conventional spherical coordinate system for many uses, in tectonics it is often unrelated to the relative motions of interest. Instead, one often benefits from using a tectonic graticule defined relative to the axis of rotation for rigid-body motion between two adjacent plates (or between a plate and the mantle). In a tectonic coordinate system, rigid-body flow for one of the two defining plates is always parallel to lines of tectonic latitude and follows small circles over the surface of Earth. However, where flow of a third plate is also important or where flow is nonrigid or non-plate like, then even the utility of a single special tectonic coordinate system can degrade.

In these general circumstances, it is thus justified to use a moving Euclidean reference frame tangent to the non-Euclidean geometry (or manifold) of interest (Cartan 1938). The idea is that vector displacements and velocities are most simple to express in Euclidean coordinates and are quite accurate when expressed in the Euclidean space tangent to a specific point of interest. If further, it is allowed to move or rotate this Euclidean space along the underlying geometry (or manifold) of interest, then displacements and velocities calculated at one point along a common flow line can be directly compared with displacements and velocities calculated at other points along the same flow line. In this spirit, I define a doubly-tangent, moving Euclidean tectonic reference frame for the present analysis. In the following paragraphs, I briefly explain how this reference frame is constructed and used.

The initial step is to pick a set of Euclidean axes, XYZ, in which the XY plane is tangent to an arbitrary point of interest (Fig. 3.1a). In Fig. 3.1a, a set of such Euclidean axes is shown relative to a triple junction point depicted between plates A, B, and C (i.e., at the intersection of the ab, ac, and bc boundaries between plates A, B, and C). The next step is to pick the axis of rotation for the relative motion inferred between a key plate pair which defines a set of small circle traces consistent with rigid-body motion between the two plates (i.e., flow along lines of tectonic latitude). In Fig. 3.1a, the rAC axis constraining relative motion between plates A and C is chosen; example small circle traces (or lines of tectonic latitude) are shown in dotted lines. In Fig. 3.1a, the ac boundary is shown to be parallel to these lines of tectonic latitude as well indicating that ac is a strike-slip boundary in this example. The orientation of X is then chosen to be parallel to the line of tectonic latitude at the point of interest.

An XYZ reference frame chosen as described above is doubly-tangent in the following sense. The XY subspace is tangent to the sphere at the point of interest, which can be shown by extending an earth-centered XZ plane out of the sphere through the point of interest (Fig. 3.1b); this construction also shows that great circle intersections of earth-centered planes radial about the point of interest project as straight lines on the given XY plane. The XZ subspace is also tangent to the relative motion between plates A and C at the point of interest, which can be shown by extending an earth-centered cone out of the sphere through the point of interest (Fig. 3.1c); this construction also shows that small circle intersections of earth-centered cones radial about the point of interest project as parabolas on the given XY plane. The projected great circle (straight line) and small circle (parabola) are shown on the XY plane together in Fig. 3.1d where the double tangency of the XYZ reference frame is explicit. In general, such a Euclidean reference frame provides a convenient context in which to evaluate the compatibility of a velocity field at the point of interest, which is the case in the triple junction stability analysis discussed further below.

As just explained, the given Euclidean reference frame may be considered (initially) static in two ways. First, the origin of the system (i.e., where the XYZ axes intersect) is fixed to the initial point of interest. As well, the attitude of the XYZ system is fixed to be tangent to this point of interest as well. To fully enable the convenience of the reference frame in a movable sense, it is critical to understand how both of these static aspects can be made to be movable (Cartan 1938).

Fig. 3.1e shows how the origin of the system can be moved along the parabola corresponding to the rigid-body tectonic motion relative to rAC and Fig. 3.1f shows the origin of the system can be moved along an arbitrary nonrigid-body path that deviates from a parabolic or small circle trajectory. In Fig. 3.1e and f, the attitude

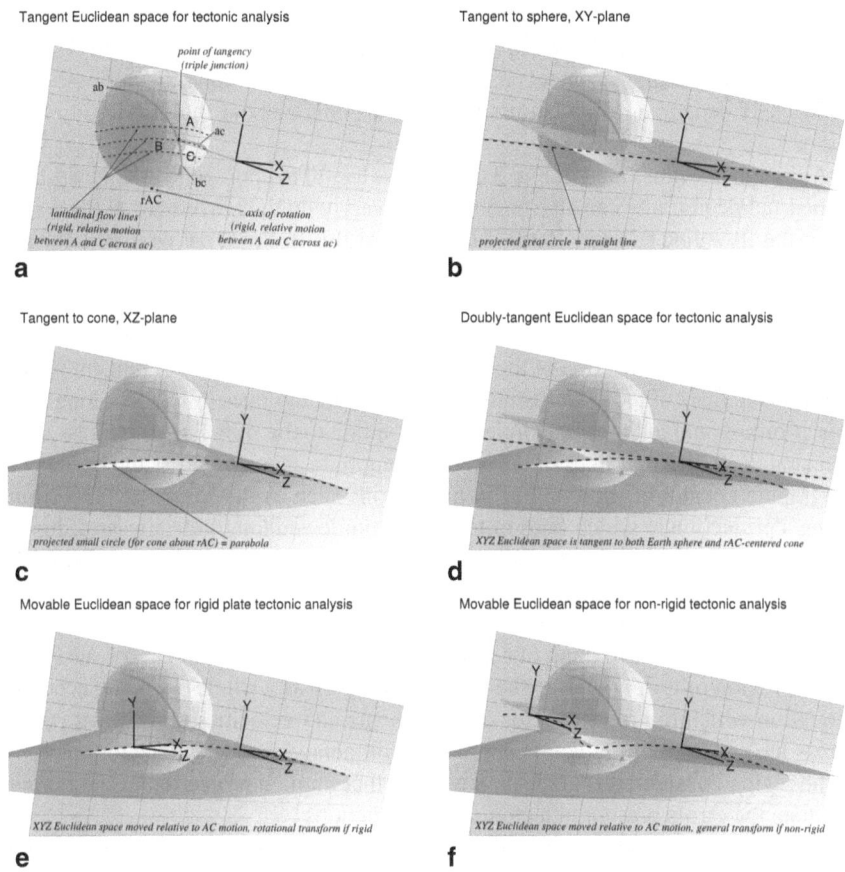

Fig. 3.1 Selection of a doubly-tangent moving Euclidean reference frame for triple junction stability analysis and displacement and velocity comparisons along common flow lines. **a** An example three plate system (A, B, and C) with three plate boundaries (ab, ac, bc) intersecting at a point. A rigid-body axis rAC is given which constrains motion of A relative to C parallel to ac. A Euclidean XYZ reference frame is constructed tangent to the Earth sphere at the triple point. **b** Projection of Earth-centered plane shows great circles project as straight lines on the XY plane. **c** Projection of Earth-centered, rAC-centered cone shows small circles project as parabolas on the XY plane. **d** XYZ reference frame is doubly-tangent; it is tangent to both Earth-centered planes and rAC-centered cones. **e** Origin of XYZ frame can be moved along rigid-body flow lines to compare displacement and velocity vectors calculated at different points along the rigid-body flow line. **f** Origin of XYZ frame can be moved along nonrigid-body flow lines to compare displacement and velocity vectors calculated at different points along the nonrigid-body flow line

of the Euclidean frame is kept fixed to be tangent to the initial point of interest (for illustration purposes) while only the origin is moved. This kind of movement is all that is necessary to enable the comparison of displacement or velocity vectors at multiple points on a common flow line. For example, if a relative displacement of ca. 1000 km is identified between plates A and C across the ac boundary at the initial point of interest, then a relative displacement of ca. 1000 km must also be identified at any other point along ac assuming that A and C are both rigid (Keppie 2012); compatible displacements must likewise be inferred along the nonrigid path in Fig. 3.1f depending on what assumptions are made about the nonrigidity of A and/or C.

Although not illustrated in Fig. 3.1, it is also useful to appreciate that the attitude of the Euclidean frame can also be allowed to vary with the movement of the origin along the flow line (Cartan 1938). The effect of this in Fig. 3.1e would be for the moving XY plane to follow a cone about the rAC axis that is tangent to the spherical Earth along the flow line of interest (i.e., a cone centered about the rAC axis that would appear as an ice cream cone holding the spherical Earth). Ripping the cone open to form a flat projection of the surface of Earth (on the movable tangent plane) would make lines of tectonic latitude appear straight and lines of tectonic longitude appear orthogonal to the small circle corresponding to the flow line of interest. This is analogous to the way lines of geographic longitude appear to cross the horizontal geographic equator at right angles in projections such as the Mercator projection. Using a projection in which lines of tectonic latitude and longitude appear orthogonal and straight relative to a given rigid-body flow line further helps to simplify the presentation and discussion of relative velocities in a Euclidean sense. If the doubly-tangent, moving reference frame just described is conceptually uncomfortable, it is also possible to think of a Euclidean reference frame tangent (or secant) to a small region of Earth (such as the Caribbean region as a whole) as being accurate enough in which to discuss vector displacements and velocities at different points across the region. Formally, this approximation perspective is satisfactory as long as the region considered is kept small enough that implied projection errors remain less than the errors present in tectonic reconstructions.

Fig. 3.2 illustrates the results of projecting the Caribbean region first in terms of a Mercator projection in conventional geographic coordinates (Fig. 3.2a), and then in terms of an oblique Mercator projection of tectonic coordinates relative to the net relative motion axis inferred for motion between North America (NA) and the Caribbean (Ca) since ca. 47.9 Ma (Fig. 3.2b). Note that the geographic graticule is an orthogonal grid in Fig. 3.2a, but the moving tectonic graticule is orthogonal in Fig. 3.2b. The moving tectonic graticule presented in Fig. 3.2b simplifies the discussion and comparison of vector (i.e., Euclidean) displacements and velocities across the Caribbean region as a whole and is thus suited to the consideration of typical geological data.

3.3 Step 2: Determine Compatibility of Major Plate Motions and Plate Boundary Orientations

To determine the compatibility of major plate motions and the general plate boundary orientations between these major plates, it is essential to remove local complexities represented by distributed and heterogeneous deformation observed across

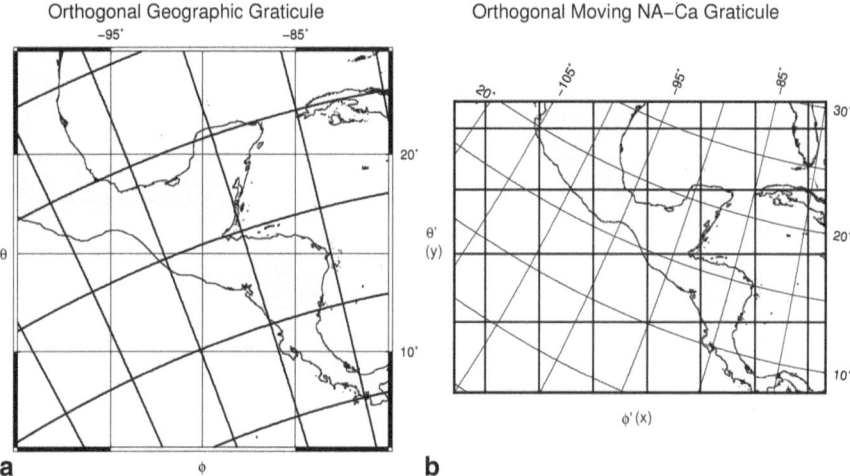

Fig. 3.2 a Geographic and tectonic graticules plotted in a Mercator projection relative to the geographic pole. The geographic graticule appears orthogonal in this case. **b** Geographic and tectonic graticules plotted in an oblique-Mercator projection relative to the tectonic pole. The tectonic graticule appears orthogonal in this case. Tectonic graticule defined relative to the net motion inferred between North America and Caribbean plates since ca. 47.9 Ma. (Leroy et al. 2000; using poles supplied by the UTIG Plates Project)

the plate boundary zones. Figure 3.3a shows the relatively complex, northern Caribbean Plate boundary zone in the NA-Ca tectonic reference frame introduced above. Section A-B in Fig. 3.3a crosses both the Polochic and Motogua fault zones (Guzman-Speziale et al. 1989), which define a microplate sliver between them, and the western Chortis rift province (Rogers and Mann 2007) which appears to show greater rifting in the north and the south. For a moment in time when deformation across all of these features is assumed to be active, the implied variation in the x-component of surface velocity, i.e., vx, from north to south along the A-B section is illustrated qualitatively in the lower frame of Fig. 3.3a. as a dashed line. This complexity can be removed for the normalization analysis by attributing the entire net change in vx to a single infinitesimally-wide plate boundary lineament at an arbitrary position along A-B (solid line in Fig. 3.3a). The same normalization is also illustrated for section C-D in Fig. 3.3a. In this case, section C-D is shown to cross the Septentrional and Enriquilo-Plantain Garden fault zones, which defined the Gonave and Puerto Rico microplates between them (Mann 2007). The net velocity change observed across C-D can be attributed to the same hypothetical northern Caribbean Plate boundary line for the normalization analysis.

The same normalization analysis is also illustrated for the South Mexico Trench and Middle America Trench segments of the western American convergent margin. Section A-B in Fig. 3.3b crosses the zone juxtaposing the Cocos (Co) and North American (NA) plates in a direction parallel to projection latitude (i.e., the x direction). This section crosses both the South Mexico Trench and the Tula-Chapala

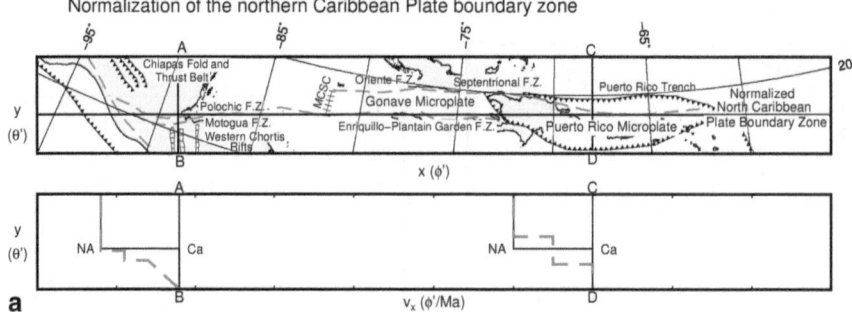

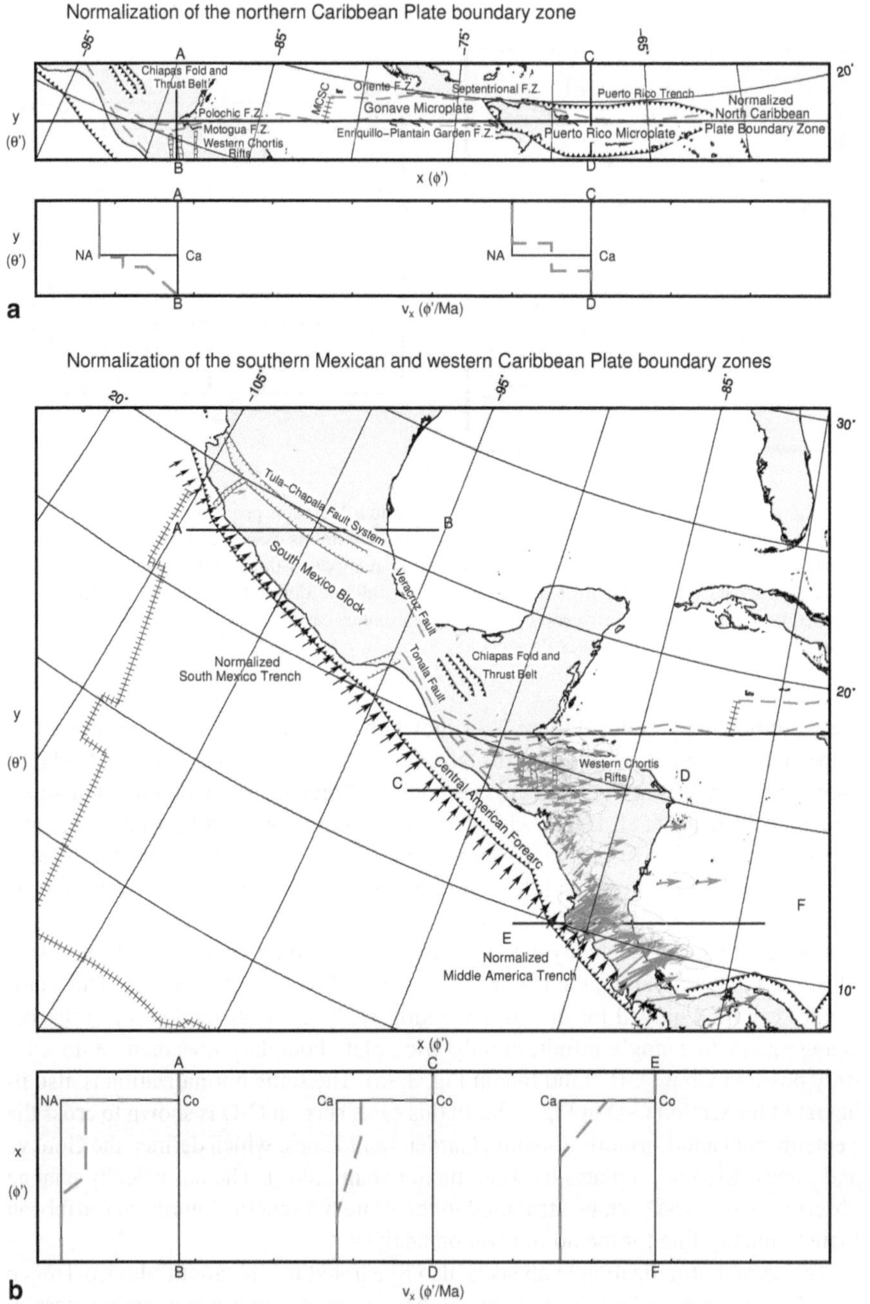

Fig. 3.3 Normalization of the North Caribbean Boundary (NCB), the South Mexico Trench (SMT), and the Middle America Trench (MAT) plate boundary zones in order to enable the identification of underlying instabilities in the major plate system. See discussion in text. **a** Normalization of the net velocity change across the NCB onto an infinitesimally-wide hypothetical lineament. **b** Normalization of the net velocity changes across the SMT and MAT onto infinitesimally-wide hypothetical lineaments. (GPS vectors in **b** from Marquez-Azua and DeMets 2009; Lyon-Caen et al. 2006; M. Rodriguez et al. 2009; and La Femina et al. 2009)

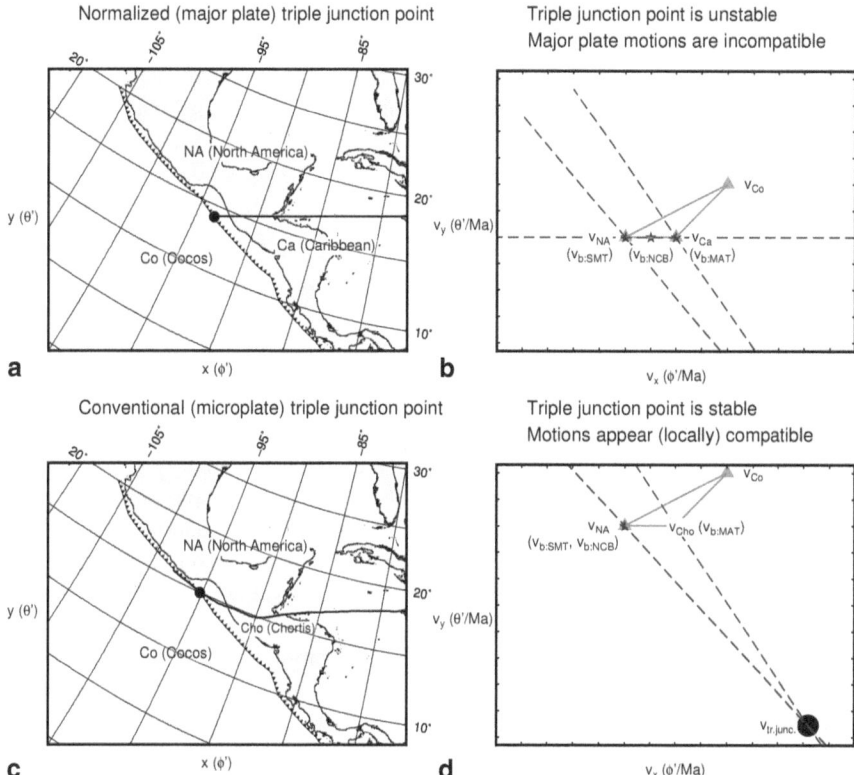

Fig. 3.4 Triple junction stability analyses for two normalized configurations of the northwest Caribbean corner zone. **a** Triple junction configuration implied by the major plate motions of North America (*NA*), Caribbean (*Ca*), and Cocos (*Co*) as normalized across infinitesimally-wide shared boundary zones (see Fig. 3.3). **b** Demonstration in relative velocity space that the triple junction configuration in **a** is unstable. See discussion in text. **c** Triple junction configuration hypothesized for the northwest Caribbean corner zone if the North Caribbean Boundary (*NCB*) is allowed to curve along a non-rigid path to become parallel to the South Mexico Trench (*SMT*). Demonstration in relative velocity space that the triple junction configuration in Fig. 3.4c is stable, at least until the moving triple point reaches the inferred bend in the NCB at which point this system will become unstable as depicted in Fig. 3.4a

Fault System, which define the South Mexico Block between them (Andreani et al. 2008). The net change in vx between Cocos and North America proper is then attributed entirely to the South Mexico Trench for the normalization analysis. Sections C-D and E-F in Fig. 3.3b cross the variably-complex zone of nonrigid deformation observed in GPS data across the western Caribbean region. The net change in the vx component of velocity between Cocos and the Caribbean proper can be attributed entirely to the Middle America Trench for the normalization analysis.

With all three plate boundaries normalized, their hypothetical intersection can then be identified by extending them to their common point of incidence (Fig. 3.4a). This configuration can then be analyzed using the classic stability analysis of McKenzie and Morgan (1969) discussed next (e.g., Keppie 2013). If the hypothetical intersection is unstable, then the non-rigid deformation removed during normalization may be a consequence of the system responding to the underlying instability just identified (e.g., Keppie 2012, 2013). If the configuration is stable, then the non-rigid deformation observed across the three plate boundaries may need to be reconciled only in terms of local changes in boundary conditions (e.g., Andreani et al. 2008; Authemayou et al. 2011).

3.4   Step 3: Determine the Stability of the Normalized, Hypothetical Triple Junction

A plate tectonic system is entirely stable if unique surface velocities can be calculated for all points on the surface of Earth. The stability analysis of McKenzie and Morgan (1969) exploits this principle. Briefly, Euclidean velocity vectors for the relative motion of each plate can be taken at their common triple point and plotted in the relative velocity space tangent to this point (grey triangles, Fig. 3.4b). Then, relative velocities for points along each plate boundary zone can be inferred from the symmetry of these boundary zones (dark stars, Fig. 3.4b). For example, components of rifting are typically assumed to be symmetric at plate boundaries, thus an intermediate velocity component to the separating plate is generally chosen. Similarly, the lower plate is typically assumed to subduct entirely under the upper plate at convergent plate boundaries, thus the trench velocity is generally assumed to be the same as the upper plate. The velocity for a stable triple point can then be deduced as follows. It is the intersection of lines in relative velocity space drawn parallel to the respective plate boundaries incident through their inferred respective velocities (dashed lines, Fig. 3.4b). A triple point is unstable if no such intersection exists (e.g., Fig. 3.4b), since this means no unique velocity can be assumed to the triple point itself.

Figures 3.4a, b show this analysis for the normalized northwest Caribbean Plate corner. Relative velocities for NA, Ca, and Co plates (i.e., vNA, vCa, and vCo) are estimated qualitatively and plotted in relative velocity space. Velocity magnitudes and exact directions are not critical since it is the relationships between the relevant velocities and plate boundaries that are important in evaluating stability. In Fig. 3.4b, velocities for NA, Ca, and Co are plotted as grey triangles and velocities for the North Caribbean Boundary (NCB), South Mexico Trench (SMT), and Middle America Trench (MAT) are plotted as dark stars. The dashed lines drawn to deduce the velocity of a stable triple point do not cross, thus we may conclude that the triple junction configuration in Fig. 3.4a is unstable.

For reference, if the NCB is made parallel to the SMT where they intersect the MAT, then the configuration is stable (e.g., Pindell and Dewey 1982), as shown in Figs. 3.4c, d. However, the velocity of the stable triple junction in this case (black circle, Fig. 3.4d) is to the southeast relative to both NA and Ca. This means it migrates down the NA coast through time as the northwest Ca region subducts under NA. Evolution of such a system will reproduce the instability identified in

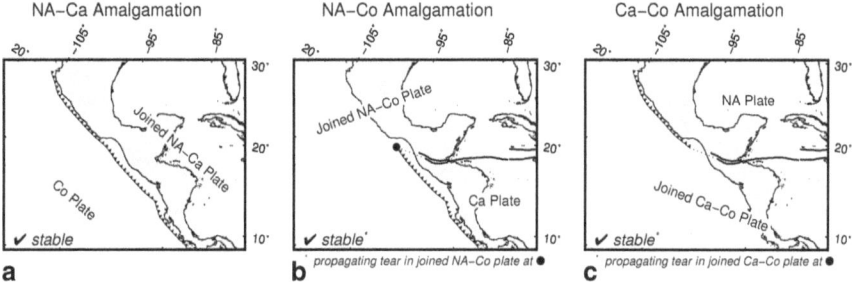

Fig. 3.5 Possible plate amalgamation solutions for the unstable northwest Caribbean Plate corner zone depicted in Fig. 3.4a: (**a**) North America (*NA*)-Caribbean (*Ca*) amalgamation, (**b**) Cocos (*Co*)-North America (*NA*) amalgamation, and (**c**) Cocos (*Co*)-Caribbean (*Ca*) amalgamation. All solutions are stable. See discussion in text

Figs. 3.4a, b when the triple point arrives at the imposed bend in the NCB (Fig. 3.4c) at which point the configuration is identical to the unstable one in Fig. 3.4a.

3.5   Step 4: Identify End-Member Nonrigid Solutions for an Unstable Triple Junction

In previous studies, including the initial one of McKenzie and Morgan (1969), the analysis of triple junction stability was primarily applied to showing when triple junctions can be stable and therefore when they can be valid parts of stable plate tectonic systems. The normalization analysis presented here opens up a further application for this type of analysis: one can construct the ways a tectonic system containing unstable triple junctions can evolve into the future (e.g., Keppie 2012, 2013). As noted above, the general ways an unstable triple junction can be resolved through further change include the following general processes: (1) plate amalgamation events, (2) plate breakup events, (3) plate velocity change events, and (4) lateral tectonic events. I now illustrate these end-member solutions in the context of the northwest Caribbean example.

Three plate amalgamation events may be imagined in the northwest Caribbean case (Fig. 3.5). If NA and Ca join (Fig. 3.5a), then the NCB vanishes which removes the main cause of the instability (i.e., a strike-slip shear zone in upper plate lithosphere striking at an angle to a lower plate). Likewise, if NA and Co join (Fig. 3.5b) or Ca and Co join (Fig. 3.5c), then the system is made stable as well. If NA and Co join, the system is stable assuming Co lithosphere continues to tear to allow rollback of the MAT with the velocity of NA (Fig. 3.5b). If Ca and Co join, the system is stable assuming Co lithosphere continues to tear to allow rollback of the SMT with the velocity of NA (Fig. 3.5c).

Three plate breakup events may be imagined in the northwest Caribbean case as well (Fig. 3.6). If Co tears (Fig. 3.6a), NA can propagate past Ca (Fig. 3.6a). In

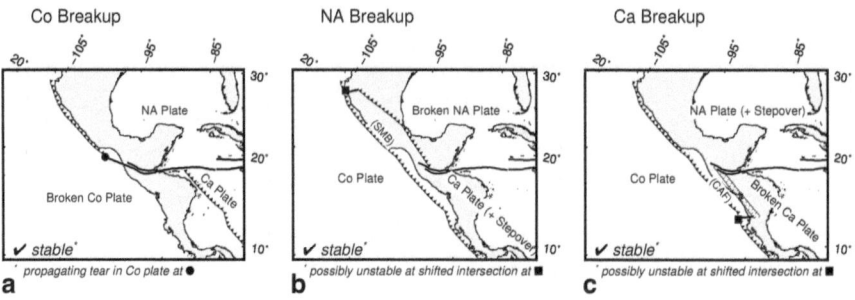

Fig. 3.6 Possible plate breakup solutions for the unstable northwest Caribbean Plate corner zone depicted in Fig. 3.4a: **a** Cocos (*Co*) breakup, **b** North America (*NA*) breakup, and **c** Cocos (*Co*) breakup. All solutions are stable, in principle. North America and Cocos breakup solutions require unidentified plate boundary changes away from the northwest Caribbean corner along the South Mexico Trench (*SMT*) and Middle America Trench (*MAT*), respectively. See discussion in text

this case, the SMT and MAT must also become segmented and Co lithosphere must continue to tear to allow rollback of the SMT with the velocity of NA and rollback of the MAT with the velocity of Ca instead. Na can also be dismembered across a convergent stepover that links the South Mexico Block (SMB) to Ca (Fig. 3.6b), or Ca can be dismembered across a divergent stepover that links the Central American Forearc (CAF) to NA. For these latter solutions, however, the unstable point is only displaced along the western American convergent margin, so these solutions can be short-lived only (occupying the time it takes to propagate the deformation) or require the presence of a boundary condition change along this margin (such as the presence of a second triple junction).

A plate velocity change between NA and Ca from pure latitudinal motion (in the NA-Ca tectonic reference frame) to a NW-SE motion can also stabilize the northwest Caribbean Plate corner (Fig. 3.7a). In general, there is a family of solutions depending on how rift symmetry across the now trans-tensional NCB is attributed (Fig. 3.7b). In Fig. 3.7a, one member of this solution family is illustrated. If Ca rifts away from a relatively fixed NA, then motion of CA parallel to the MAT can stabilize the northwest Caribbean corner. In Fig. 3.7b, the family of solutions is illustrated. The other end-member solution is if NA rifts away from a relatively fixed Ca. In this case, motion of NA parallel to the SMT can stabilize the northwest Caribbean corner. Intermediate solutions lie between these two end-members. For convenience, I adopt a convention to represent this family of solutions as follows (see Fig. 3.7d). I select a specific boundary (in this case the NCB) and a specific plate (in this case the Ca) with respect to which I specify the family of stable solutions by specifying the family of stable relative velocities possible for the selected plate. I draw a line orthogonal to the specific boundary (i.e., the NCB) in relative velocity space to represent all possible velocities for the specific plate (i.e., Ca). I then overlay a thick grey line segment on this line to show those velocities which are stable relative to the hypothetical triple junction (Fig. 3.7b).

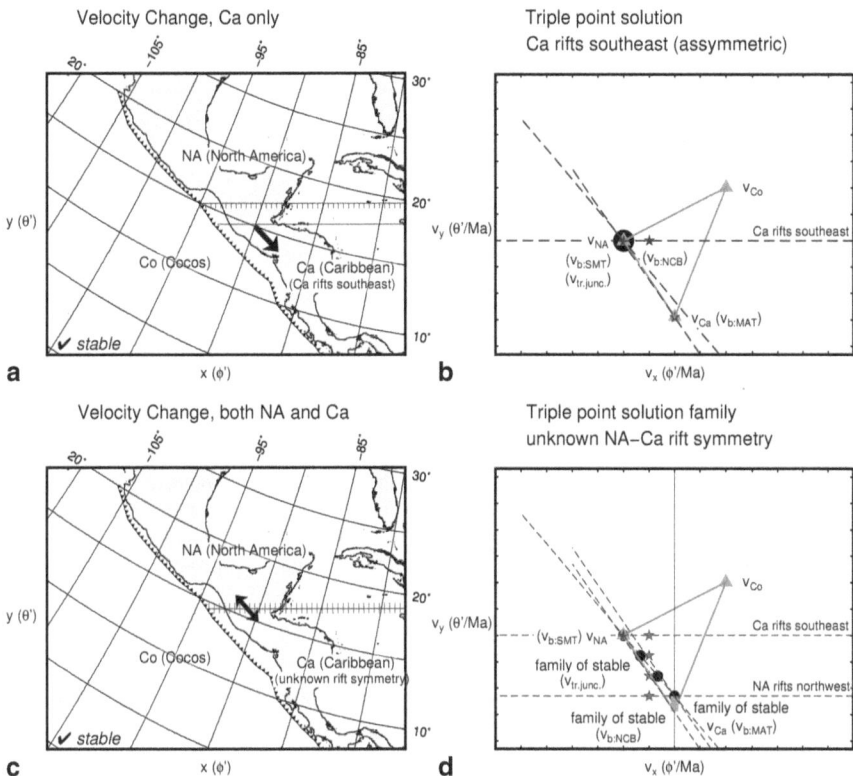

Fig. 3.7 Possible plate velocity change solutions for the unstable northwest Caribbean Plate corner zone depicted in Fig. 3.4a. If the NCB becomes appropriately trans-tensional, the system is stable. **a** Ca rifting away from NA parallel to the MAT could stabilize the northwest Caribbean corner. **b** Ca rifting away from NA parallel to the MAT, NA rifting away from Ca parallel to the SMT, or mutual rifting of NA away from Ca and vice versa parallel to an azimuth intermediate to those of the MAT and SMT could stabilize the northwest Caribbean plate corner. See discussion in text

Lateral tectonics can also stabilize the northwest Caribbean corner (Keppie 2012, 2013). If trench rollback for both the SMT and MAT segments is equal, then relative motion of NA west past Ca opens up an apparent divergent zone in the western part of Ca. Instead of this zone being filled by vertically-intrusive magmas, for example, it can also be filled by laterally-intrusive microplates captured from the southern part of NA. This solution resembles the upper plate breakup solutions noted above but has important differences. Principally, in pure form, it can resolve the instability even when there are no plate boundary changes in the system to the north or south of the northwest Caribbean (Fig. 3.8a). In this case, microplates must be captured out of the interior of NA, i.e., from the western Gulf of Mexico region (Keppie 2012). If, however, a plate boundary change is present in the system to the north of the northwest Caribbean, a modified version of lateral intrusion can also take place (Fig. 3.8b). In this case, microplates can be captured from the exterior part of NA,

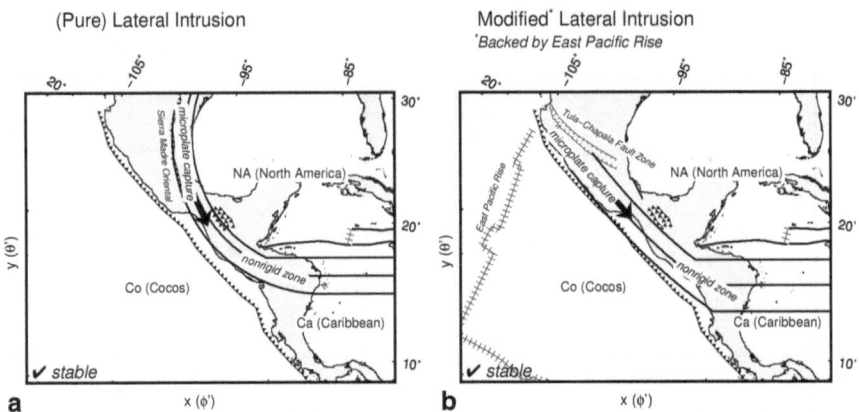

Fig. 3.8 Possible lateral tectonic solutions for the unstable northwest Caribbean Plate corner zone depicted in Fig. 3.4a. **a** Microplates captured from the interior part of southern North America could stabilize the northwest Caribbean corner. **b** Microplates captured from the exterior part of southern North America could stabilize the northwest Caribbean corner, if a plate boundary condition change takes place somewhere to the northwest along the South Mexico Trench (*SMT*). Subduction of the East Pacific Rise could be an appropriate boundary condition change along the SMT. See discussion in text

at least until the deflected NCB reaches the point of boundary condition change. In Fig. 3.8b, the subduction of the East Pacific Rise may provide a boundary condition change along the SMT sufficient to facilitate the modified version of lateral intrusion since ca. 10–6 Ma in its present position and back to ca. 40–30 Ma in positions farther north. Prior to the Oligocene, however, it is not clear that a modified version of lateral intrusion could apply in the northwest Caribbean case. These issues are considered more fully in the main paper.

References

Andreani L, Le Pichon X, Rangin C, Martínez-Reyes J (2008) The southern Mexico block: main boundaries and new estimation for its Quaternary motion. B Soc Geol Fr 179(2):209–223

Authemayou C, Brocard G, Teyssier C, Simon-Labric T, Guttiérrez A, Chiquín EN, Morán S (2011) The Caribbean–North America–Cocos Triple Junction and the dynamics of the Polochic–Motagua fault systems: pull-up and zipper models. Tectonics 30:TC 3010. doi:10.1029/2010TC002814

Cartan E (1938) La theorie des groupes finis et continus et la geometrie differentielle traitees par la methode du repere mobile. B Am Math Soc p 269

Cloetingh S, Negendank J (eds) (2010) New frontiers in integrated solid earth sciences, International Year of Planet Earth Series XIX. Springer Dordrecht, Heidelberg, p 414

Guzmán-Speziale M, Pennington WD, Matumoto T (1989) The triple junction of the North America, Cocos, and Caribbean plates: seismicity and tectonics. Tectonics 8(5):981–997

Keppie DF (2012) Derivation of the Chortis and Chiapas blocks from the western Gulf of Mexico in the latest Cretaceous–Cenozoic: the Pirate model. Int Geol Rev 54(15):1765–1775. doi:10.1080/00206814.2012.676356

Keppie DF (2013) The rationale and essential elements for the new Pirate model of Caribbean tectonics. Geoscience Canada 40(1). doi: http://dx.doi.org/10.12789%2Fgeocanj%2F2013.40.002

La Femina PC, Dixon TH, Govers R, Norabuena E, Turner H, Saballos A, Mattioli G, Protti M, Strauch W (2009) Fore-arc motion and Cocos Ridge collision in Central America. Geochem Geophys Geosys 10(5). doi:10.1029/2008GC002181

Lyon-Caen H et al (2006) Kinematics of the North American-Caribbean-Cocos plates in Central America from new GPS measurements across the Polochic-Motagua fault system. Geophys Res Lett 33:5

Mann P (2007) Overview of the tectonic history of northern Central America. Geol Soc Am 428:1–19 (Special Paper)

Marquez-Azua B, DeMets C (2009) Deformation of Mexico from continuous GPS from 1993 to 2008. Geochem Geophys Geosys 10:16

McKenzie DP, Morgan WJ (1969) Evolution of triple junctions. Nature 224(5215):125–133

Pindell JL, Dewey JF (1982) Permo-Triassic reconstruction of western Pangea and the evolution of the Gulf of Mexico/Caribbean region. Tectonics 1(2):179–211

Rodriguez M, DeMets C, Rogers R, Tenorio C, Hernandez D (2009) A GPS and modelling study of deformation in northern Central America. Geophys J Int 178(3):1733–1754

Rogers RD, Mann P (2007) Transtensional deformation of the western Caribbean–North America plate boundary zone. Geol Soc Am 428:37–64 (Special Paper)

Seton M, Müller RD, Zahirovic S, Gaina C, Torsvik TH, Shephard G, Talsma A, Gurnis M, Turner M, Maus S, Chandler M (2012) Global continental and ocean basin reconstructions since 200 Ma. Earth-Sci Rev 113(3–4):212–270